Introduction Chemical Engineering

S. Pushpavanam

Professor
Chemical Engineering Department
Indian Institute of Technology Madras

PHI Learning Private Limited

Delhi-110092
2017

₹ 195.00

INTRODUCTION TO CHEMICAL ENGINEERING
S. Pushpavanam

© 2012 by PHI Learning Private Limited, Delhi. All rights reserved. No part of this book may be reproduced in any form, by mimeograph or any other means, without permission in writing from the publisher.

ISBN-978-81-203-4577-5

The export rights of this book are vested solely with the publisher.

Third Printing ... ... **January, 2017**

Published by Asoke K. Ghosh, PHI Learning Private Limited, Rimjhim House, 111, Patparganj Industrial Estate, Delhi-110092 and Printed by Raj Press, New Delhi-110012.

Contents

Preface   vii

1. Role of a Chemical Engineer **1–26**

Introduction   *1*
Chemical Engineering in Everyday Life   *1*
Scaling Up or Down   *3*
Engineering Application of Portable Devices   *5*
Challenges in the Petroleum Sector   *6*
    Transport Across the Ocean Bed   *6*
    Operations in a Refinery   *7*
    The Language of the Refiner   *12*
    Coal Gasification   *13*
Euro Norms/Bharat Stage Norms to Curb Atmospheric Pollution   *15*
    Versatility of a Chemical Engineer   *18*
    Role of Chemical Engineers in Biomedical Engineering   *22*
    Similarities in Dissimilar Applications   *23*
Exercises   26

2. Modern Chemical Engineering Plants **27–53**

Introduction   *27*
Batch Processing   *28*
Paint Manufacture   *31*
The Transition from Batch to Continuous Processing   *31*
Case Study 1: Manufacture of Sulphuric Acid   *34*
    Lead Chamber Process   *34*
    Recycle of NO   *35*
    Contact Process   *39*
Implications of Coupling and Recycling: Start-up and Shutdown   *41*
Case Study 2: Soda Ash (Sodium Carbonate) Industry   *43*
    Leblanc Process   *43*
    Solvay Process   *43*

Common Features between the Evolution of the Sulphuric Acid Industry
and the Soda Ash Industry   *45*
Processes with Recycle Streams   *45*
Reverse Osmosis Plants   *46*
Typical Processes Involved in the Plant   *48*
   Rapid Gravity Filter   *48*
   Disinfection   *49*
   Reverse Osmosis   *49*
Description of the Process of RO Plant in IIT Madras   *49*
Exercises   *51*
APPENDIX—Liquid Nitrogen Plant   *52*

3. Chemical Engineer and Chemical Engineering Profession 54–72

Introduction   *54*
The Birth of Chemical Engineering   *55*
Distillation   *56*
Curriculum   *58*
   Thermodynamics   *59*
   Momentum Transfer   *60*
   Heat Transfer   *61*
   Mass Transfer   *64*
   Chemical Reaction Engineering   *69*
Remarks   *72*

4. Role and Importance of Basic Sciences in Engineering 73–105

Introduction   *73*
An Application of Conservation of Mass in a Closed System   *74*
A Practical Application of this Idea to the Source
   Apportionment Problem in Air Pollution   *77*
Conservation Laws, Closed Systems and Open Systems   *82*
Infinitesimal Control Volume   *85*
Macroscopic Control Volume   *87*
   Unsteady State: Ordinary Differential Equations   *87*
   Stability Issues in a System   *90*
Conservation of Mass: Applications   *92*
Conservation of Mass Reacting Systems   *93*
Application to a Flash Unit: Nonlinear Algebraic Equations   *95*
Physically Admissible Solutions   *98*
An Example Showing a Partial Differential Equation   *100*
Optimization Problems in Chemical Engineering   *103*
Exercises   *105*

Contents v

5. **Dimensionless Analysis and Scale-up:** Another Illustration of How Physics and Mathematics can be Combined **106–125**

   Introduction   *106*
   Safety Issues in the Scale-up   *107*
   Lab Scale and Commercial Scale   *108*
   Dimensionless Analysis: Dimensionless Numbers   *111*
   Obtaining the Friction Factor   *118*
   Application of Dimensionless Analysis to a Reactor, i.e. A CSTR   *122*
   Conclusions   *124*
   Exercises   125
   Reference   125

6. **Semi-empirical Approach in Engineering:** Departure from Scientific Rigor—Applications in Atmospheric Pollution and Turbulence **126–136**

   Introduction   *126*
   Motivation for Semi-empirical Approach   *127*
   Atmospheric Pollution and Dispersion   *128*
   Applications in Turbulent Flows   *134*
   Exercises   136

7. **Safety, Health, Environment and Ethics** **137–166**

   Introduction   *137*
   Safety in Chemical Process Industries   *137*
      Lessons for the Management   *140*
      Importance of Quantitative Information   *141*
   Case Study 1: Extinction of Different Species of Vultures   *142*
      Main Reasons for Vulture Deaths   *144*
      Steps That Need To Be Taken   *145*
   Case Study 2: DDT   *145*
      The Early Years   *145*
      How Does DDT Work as an Insecticide?   *146*
      Advantages of Using DDT   *146*
      Health Effects   *147*
      Residues in Food   *147*
      Ecological Effects   *148*
      DDT Restrictions   *148*
   Case Study 3: Environmental Hazards of a Green Project   *149*
      The Adverse Effects of this Green Project   *150*
   Case Study 4: Endosulfan   *151*
      Effects of Aerial Spraying of Endosulfan on Residents of Kasaragod, Kerala   *151*
      Effect of Endosulfan use in Mango Orchards on Residents of Palakkad District, Kerala   *152*
      Conflict between States and Central Government   *153*
      Impact of Endosulfan   *153*

Case Study 5: Plachimada Bottling Plant of Coca-Cola *156*
Case Study 6: Marine Disasters in the Arabian Sea Near Mumbai *158*
Ethics *159*
Case Study 7 *159*
Case Study 8 *160*
Case Study 9 *161*
Case Study 10 *161*
Case Study 11 *162*
Case Study 12 *162*
Exercises 164
References 165

Appendix—MATLAB Programs *167–170*
Index *171–173*

Preface

This book is an outgrowth of my teaching the course on *Introduction to Chemical Engineering* (CH1010) for the first time to the fresh undergraduate students of IIT Madras in the July–December 2009 semester. Never having taught a fresh class before (in a career of twenty years), I was impressed to see the students looking young, fresh, motivated, and eager to learn. Attendance in the class was close to a 100% and this was in itself something I, as an instructor, could look forward to. It was nice to experience the eagerness and the enthusiasm of the students to learn and experiment. There was a healthy interaction during all the lectures, with students putting questions, which was a positive sign. The course is extremely important as it sets the tone in the minds of the students about the entire four-year degree programme. Getting the right instructor and the right set of course contents can get the students a good feeling about the programme they are in and can make them stay motivated throughout the programme. It can be the determining force as to whether the students are attracted to or repelled from their major discipline "chemical engineering".

There was an intense discussion amongst the faculty of IIT Madras on the modes and methods for teaching this course during the January–March 2009 semester. One point of view was to have different faculty exposing the students to the different areas of chemical engineering. Here each faculty would introduce the students to his research area in a couple of lectures. The proponents of this approach were of the opinion that this would help match the aptitudes of students to different research areas in chemical engineering. The drawback here would be that the students were just out of school and would not be mature enough and ready for exposure to research challenges. Administratively, there would be no single person accountable for the quality of the course. Moreover, there would be an inherent discontinuity in this approach with different research areas being covered over short intervals of time. The differences in the style of teaching would also result in a course which is not cohesive. The frequent changes in the style would leave the students confused. Subsequently, it was felt that adopting this approach would not be fair to the students.

Being the first course in which the students would be exposed to the department it was necessary that the person teaching the course keep in mind that he must make the discipline attractive and informative. To be effective the course had to be designed keeping in mind the intellectual and the social backgrounds of the students and their maturity levels. The final shape of this course and the contents of this book have been strongly influenced by Professor Krishnaiah, my teacher and guru, as well as Niket Kaisare, a young colleague of mine. I think if I had not agreed to teach the course the arguments amongst my faculty colleagues could have been continuing even today.

The contents of the course have been developed keeping in mind the educational background of the (10+2) school students who upon joining the engineering education would like to know the relevance of their school education to the chemical engineering profession. The book is therefore aimed primarily at the first-year students of engineering, who already have a background in the basic sciences: physics, chemistry, and mathematics. They are trained to solve problems in these subjects and are trained to think that all problems have unique answers.

One of the objectives of the book is also to introduce the students to lateral thinking so that they realize that a problem may have different solutions. The examples and assignments in this book have been chosen to drive home the point that solutions to a problem in the real world do not have to be unique. The emphasis is to show how common sense combined with a strong fundamental scientific knowledge is the basic quality necessary to be a good engineer. This would help an engineer find innovative solutions.

The book explains the issues involved in the chemical engineering discipline and shows how the concepts of physics, chemistry, biology, and mathematics are used in solving them. In addition to this, there is also a section on ethics to help students understand their role in society. Most of the case studies for the ethics section are from realistic but hypothetical situations and have been chosen such that the students identify themselves with them.

The examples and different topics covered in the book have been arrived at on the basis of informal discussions with several colleagues in the corridors of IIT Madras. As a part of this course, students were taken to the water treatment plant in the swimming pool, the reverse osmosis (RO) plant which provides drinking water to the students and the liquid nitrogen plant. All these are located in the institute. This made them understand the level of details an engineer has to be concerned with when it comes to designing and running a process plant. These plants have backup features in case certain units fail, and features such as these can be best appreciated only during a site visit. Similar plants would be present in other colleges and can be ideal for students to get a practical feel of the basic principles of chemical engineering.

I express my sincere thanks to my colleagues Ramanathan, Nagarajan, and Sethupathi from IIT Madras for providing valuable inputs to several sections of the book. Renganathan went through the entire manuscript with a magnifying glass and provided several critical inputs and suggestions to improve its quality. My students—Manoj Yadav, Hemalatha, Ravikanth, Varshaa Naganathan, and Ashraf Ali—helped with the figures and programs in the text. Jason, in particular, went through the entire manuscript and provided a student's perspective to help improve the contents.

The book would not have been possible without the encouragement and support of my wife, Geetha. My cousin Madhusoodhanan also played an important role by prompting me every time we met as to when the next book was coming out.

Finally, I would like to thank the Centre for Continuing Education, IIT Madras for financial support for writing this book.

S. Pushpavanam

1

Role of a Chemical Engineer

Introduction

In this chapter, we discuss different areas in which chemical engineers play a vital role. Various products manufactured by them are first described. The challenges faced by the chemical engineer and their roles in different industrial contexts are emphasized. These show that the role of chemistry in chemical engineering is minimal and there is a strong component of other scientific disciplines in chemical engineering. The diverse examples presented in this book help establish the versatility of the chemical engineer. This versatility helps the chemical engineer find employment in a wide spectrum of industries: from petroleum refining to semiconductor processing. The examples also show how the chemical engineer comes up with different applications that are based on similar underlying fundamental principles. For instance, membrane separation processes are used for applications ranging from haemodialysis to desalination.

Chemical Engineering in Everyday Life

The first question which naturally arises to the curious student is: What does a chemical engineer do? What does he manufacture? The products manufactured by chemical engineers can be found in various items used by the common man as a part of his regular routine in everyday life. Beginning with the toothpaste you use when you get up from bed, the toothbrush (bristles and the handle), the soap, the shampoo, the hair oil or shaving cream or gel one can think of numerous products made by chemical engineers which are an integral part of our daily life. Similarly, the detergent used for washing clothes, the petrol or diesel in your vehicle, the ink in your pen, and the paper you use to write on are all products manufactured by chemical engineers. Most of us use them on a regular basis without thinking about how they are made and how the composition of each product is determined. CDs and different electronic chips used in computers and other electronic gadgets are all manufactured using chemical engineering principles. The chemical engineer is concerned not only with the manufacture of these compounds but also improving their quality by ensuring that they have desired physical and chemical properties. These are usually dictated by market demands and other considerations such as impact on health and environment. Hence, chemical engineers ask themselves what are the properties required of the end product they are manufacturing. This helps them design new processes to improve the quality of existing products. For instance, *liquefied petroleum gas* (LPG), used for domestic cooking in common households, is essentially a mixture of propane and butane. This is a "clean" fuel made by chemical engineers as it does not adversely impact the environment. When using this fuel combustion is complete and no unconverted hydrocarbons are released into the atmosphere. As compared to the kerosene stoves and coal-based stoves used in the past, this does not have any adverse effect on indoor air quality and, hence, the health of individuals in the household. Combustion is incomplete in kerosene-based stoves resulting in the presence of unburnt hydrocarbons in the indoor air. The shift to LPG has improved fuel efficiency as well as indoor air quality. Several state governments have policies facilitating and encouraging the use of LPG amongst the poor sections of society. This reduces their dependence on firewood and helps in maintaining a green environment.

In addition to the products mentioned so far in the domestic context, chemical engineers are also actively involved in the fertilizer industry, pharmaceutical industry, biotechnology, bio-chemical engineering, fossil fuels (coal-based power plants and petroleum refineries where useful fractions are obtained from petroleum crude), biofuels, renewable sources of energy like solar energy and wind energy, etc.

As chemical engineers we must learn how to manufacture these products (i) in an economical way, (ii) in a safer way, and (iii) in an environmentally friendly way.

The sound knowledge of physics, chemistry, biology, and mathematics combined with engineering principles makes the chemical engineer versatile. This broad knowledge base gives the chemical engineer an opportunity to work in a wide range of industries and disciplines. This makes chemical engineering interdisciplinary. Chemical engineers always ask themselves what are the properties required of the product they are manufacturing. This forces them to innovate continuously and improve the quality of the existing products and bring about changes in technology. For instance, in the area of semiconductor processing to negate the influence of gravity effects they study crystal growth under micro-gravity conditions.

Now let's discuss some of the challenges faced by chemical engineers.

Scaling Up or Down

Typically, the processes which lead to the manufacture of an end product are developed in a laboratory. Depending on the processes involved this needs a basic knowledge of chemistry, biology, and physics. Biology comes in when a microorganism or enzyme is used to manufacture a product. The product engineer or chemist identifies a feasible route or mechanism to obtain the end product at the laboratory scale. The process engineer takes the idea developed at the laboratory and scales it up for production at the industrial scale. The transformation from a concept in the laboratory to the process in a commercial plant, i.e. increasing production from a few grams per hour to a few tons per hour is one of the challenges faced by chemical engineers. This requires the knowledge of mathematics, physics, chemistry, biology, and economics along with considerable innovation.

The scaling-up operation is usually done in several steps. After an idea is developed in a laboratory a demonstration plant of an intermediate scale or size (known as pilot plant) is built before going in for the full-scale commercial production. This helps identify new challenges that have to be addressed as a result of increasing the scale of production. The operation of the pilot plant results in a lot of new knowledge being generated. The operation at this scale allows us to test different ideas to improve the performance. For instance, in a large vessel mixing the contents to maintain a uniform composition could be a challenge. Similarly, ensuring a uniform temperature inside a large vessel is difficult as mixing is usually poor. These issues have to be resolved before going from a small-scale to a large-scale production. Pilot plants for demonstration of a technology can be significantly large in size.

The *integrated gasification combined cycle* (IGCC) process based on coal gasification is equipped with gas turbines and steam turbines. The gasifier generates a mixture of CO and H_2. This is burnt in the gas turbine and the energy present in the hot gases leaving the gas turbine is used to generate steam which in turn drives a steam turbine. A "demonstration power plant of the IGCC process "based on coal gasification and built to deliver 6 MW of power would be the size of a six-storey building spanning at least two blocks. The typical requirements of power in a city are several orders of magnitude higher. Chennai city has a demand of around 10,000 MW. This gives us an idea of the size of thermal power plants which are needed for sustaining power requirements of a city. The confidence in operating these commercial power plants stems from the experience gained in operating the smaller demo units.

Commercial plants are typically "BIG". The large size of chemical plants arises because chemical engineers exploit the economy of scale. This means a reduction in the cost of manufacture per unit quantity of the product is obtained as we increase the total production. Of course, in addition to the large scale of production, it is important to have energy-efficient practices incorporated into the design and plant operation to improve the efficiency and reduce the costs. In large plants the opportunities for this are more as the freedom to innovate, integrate, and modify is higher. There are more avenues to improve the efficiency of a plant operation. For instance, the energy released in a particular unit can be recovered in another unit via heat exchangers by a process called the *heat integration*. The chemical engineer is trained to identify the areas in a plant where improvements can be made.

More recently, chemical engineers have been involved with scaling-down processes, so that the entire process of production of a chemical or analysis of a chemical can be carried out on a small chip. An example of this is the hand-held device to measure sugar levels in the blood for patients with diabetes. This instrument gives an instantaneous reading and has made the laborious and time-consuming laboratory tests a thing of the past. However, the accuracy of the measurements from these instruments which give instantaneous values may be lower than the measurements made using the classical methods.

The shrinking in the size of these small devices leads to a reduction in the length and time scales. These devices are used in hospitals to quickly determine if sugar levels are under control in an emergency. This quick estimation can be life-saving. The major advantage of the scaling-down approach is that it affords a possibility to have a decentralized way of producing and analysing chemicals. This can be viewed as being analogous to the revolution in the computer industry where the size of computers has reduced drastically over the last few decades. As a result of this, a lot of information can now be stored in chips of increasingly smaller sizes. In the 1970s the size of computers was big and would occupy large rooms. Today we have the same computational power, if not more residing in desktops. Modern-day laptops and palm-held devices can be used to carry out the tasks of the supersized computers of the earlier days. The shrinking in the size of computers has been made possible by the decrease in the size of chips.

A similar revolution is being envisaged in chemical engineering where we aim to develop an entire lab on a chip or carry out an entire process on a chip. In the former, the interest is to analyse chemicals in a sample (detect ions in wastewater), whereas in the latter it is to manufacture a product. There are several situations in which the scale-down approach is advantageous. For instance, it helps in designing a portable system for analysing water samples in remote areas. In the design of these systems, the challenge is to carry out several processes, i.e. reactions and separations in systems whose dimensions are in the sub-millimetre or micron range. The challenge in carrying out these processes at microscales is to improve the mixing of chemicals at low flowrates to enable reaction, separation, and detection.

The ultimate goal is to see if after shrinking the size of operations it is possible to carry out several tasks or tests simultaneously. For instance, can we test the effectiveness of different catalyst compositions or formulations at the same time in an economical way? Another way to understand this is to think of the mobile phone which seems to be omnipresent these days. These phones come packed with many features: camera, alarm clock, e-mail, etc. The miniaturization of laboratory on a chip aims at a similar revolution in the chemical industry wherein it may be possible to carry out several reactions, analyse several species and produce several products in a small hand-held device. Here an advantage would be the low chemical inventories giving rise to improved safety features. The amount of sample required for analysis would also be low compared to the classical tools used.

Engineering Application of Portable Devices

There is a strong motivation for developing applications in environmental engineering using hand-held devices. The quality of water in several places has deteriorated. In most developing countries, in large areas the groundwater sources are contaminated with fluoride, arsenic, and chromium. These affect the quality of potable water in remote villages. Testing the water for these toxic compounds used to be done in analytical laboratories in big cities. However, if continuous monitoring of water quality is required there is a strong need to have a decentralized analytical testing facility. This can be achieved through a portable device or a lab on a chip.

Such a device will also be useful in cities. We now provide a motivation for this. The wastewater coming out of individual households and buildings in big cities is transported through pipes and treated at a centralized plant. The *pollutants* in this sewage water primarily consist of soap, detergent, and some organic matter which come from kitchen wastes.

Centralized treatment plants in cities are designed to operate at a particular capacity or to process a certain flowrate of waste or sewage water. With the increase in population in cities the load on the treatment plants has dramatically increased. Consequently, the treatment of the wastewater in these plants is not very effective.

One possible method to avoid this problem is to develop treatment methods which are cost effective and can be implemented at the level of an individual house or a building with several households. This approach would facilitate reuse, and recycle of treated wastewater. Alternatively, it can also be used to recharge the groundwater table locally. There is a strong impetus for R&D in this area.

Domestic water coming from household bathrooms is called the *greywater*. This contains primarily soap and detergent and hence the pollutant levels are very low. One option is to use phytoremediation or plants to treat this water and recharge the groundwater table. Identification of plants which can survive the alkaline nature of the greywater, taking into account the intermittent flowrate of the wastewater is an important challenge in this field. The discharge at discrete instant arises since this wastewater is produced only when the washing machine or the bathroom is used. This decentralized treatment at the household level can help promote sustainable development. While using this approach it is necessary to frequently analyse the groundwater levels to ensure that they are not contaminated and that the phytoremediation technique is actually effective. The level of nitrates and phosphates in the groundwater has to be periodically monitored. A lab on a chip which can do this would be very beneficial.

Hence, while the economy of scale may drive one towards large-size chemical plants, there is also a strong drive towards miniaturization of plants. To go for a large plant or a lab on a chip is determined purely by the application, economic, and safety considerations. Chemical engineers operate at these two different ends of spectrum: large plants exploiting economies of scale and micro systems such as a lab on a chip.

We now describe some industries in detail to highlight their important features. This also brings out the important challenges and issues faced by the chemical engineer and the steps taken to resolve them. In addition to the classical industries the role of chemical engineers in several esoteric areas such as semiconductor processing and biomedical engineering is discussed.

Challenges in the Petroleum Sector

Transport Across the Ocean Bed

The crude oil found in the underground reservoirs consists of a complex mixture of hydrocarbons. Civil engineers and ocean engineers (associated with offshore operations) are concerned with bringing the oil from the reservoir to the ground level. These are called *upstream operations*. As opposed to this, *downstream operations* consist of separating the crude oil into different fractions such as diesel or petrol on the basis of boiling point differences. It also consists of steps to improve the quality of these fractions using processes such as hydrodesulphurization. Here the sulphur containing compounds react catalytically with hydrogen to give hydrogen sulphide. This is an important step as it helps reduce release of sulphur oxides from automobile exhaust in atmosphere and reduces the poisoning of catalysts in the refining industry.

At the interface of the upstream and downstream processing is the transport of crude oil from the oil wells to the refinery. This is usually done through pipelines running along the ocean floor in case of an offshore installation.

The oil coming out of a well can contain long chain *n*-alkane hydrocarbons. These are typically saturated hydrocarbons with more than 20 carbon atoms. The presence of these large molecules can create problems in transportation as they are responsible for the increased viscosity of the crude oil. *Viscosity* acts like friction and slows down the flow of fluid making it difficult and energy intensive to pump the fluid. The presence of these molecules can also cause settling of wax in storage tanks, and decrease the inner diameter in pipelines, further restricting the flow. The waxy fraction which has a relatively high melting point of around 35°C has a tendency to crystallize and adhere to the surfaces of the wall in a pipeline. It gets deposited along the surface and increases wall roughness and eventually may even block the flow. This results in a higher or increasing pressure drop and reduces the flowrate.

Wax crystallization first occurs at a temperature of around 35°C. The solubility of wax is strongly dependent on temperature. In most oil reservoirs this wax is completely dissolved due to the high temperatures prevailing in them. The wax is in the dissolved state at a temperature of 60–70°C (which prevails in the reservoirs) but as the temperature drops to below 35°C during transportation it solidifies on the walls of the pipe. This is a common problem in the pipelines which run along the seabed in the North Sea. Chemical engineers are responsible for predicting the location where crystallization of wax can occur. They must also design control actions which need to be taken to avoid this. The appearance of wax is determined by how quickly the contents of the pipe lose heat to the surroundings. This determines the length at which the temperature will fall below the melting point of wax 35°C leading to its crystallization. Hence the transfer of heat from the pipeline has to be analysed to determine the onset of crystallization and the control action which needs to be taken. This involves the understanding the physics of heat transfer, i.e. what are the factors which quantitatively determine the heat transfer rate. This is an example of how heat transfer from the pipeline a physical process has an application in chemical engineering.

Operations in a Refinery

Chemical engineers are also responsible for separating crude oil from oil wells to different useful fractions. This is typically done in a refinery. Petroleum refining is a classical industry which is associated with chemical engineering. This is a downstream processing operation. Here the crude oil is separated and treated to get useful products. The basic process used for this is *distillation*. Here the difference in the boiling points of the various fractions is exploited to separate the mixture into useful products. As the crude oil is heated the most volatile components first vaporize. As the temperature is increased further the compounds with slightly higher boiling points vaporize. Since the crude oil contains a mixture of many hydrocarbons, the products are classified on the basis of the boiling point range in which they lie. The species belonging to a

8 Introduction to Chemical Engineering

particular boiling point range are classified as gasoline, kerosene, diesel, etc. The size of the molecules can be approximately represented by the number of carbon atoms they contain. In general the larger the number of carbon atoms, the more is the molecular weight, the more is the boiling point and the more viscous is the liquid. Molecules with 5–7 carbon atoms are light in the sense that they have low boiling points or a high vapour pressure. These are clear liquids which can be easily vaporized and are called *naphthas*. These have a boiling point in the range of 60–100°C. These are used as solvents-dry cleaning liquids, solvents for paints, etc.

Molecules with 7–11 carbon atoms are blended together for gasoline or petrol (boiling point 40–205°C). Kerosene has molecules in the C_{12} to C_{15} range. This is used as a fuel for jet engines and has a boiling point in the range of 175–325°C. Diesel and heavier fuel oils have C_{16}–C_{18} molecules and have a boiling point range of 250–350°C. Each fraction is defined so that it consists of compounds which have boiling point in a specific range. The composition of a fraction such as, say, diesel is hence not uniform across gas stations. The petrol fraction in different gas stations can contain a larger or a smaller amount of a particular compound. The composition of diesel can also vary in a gas station on a daily or weekly basis. Only the compounds in the boiling point range are known but their mass fractions can vary. Some of the key properties which define the petrol fraction such as the boiling point range are similar across gas stations. To reduce the variability of composition of petrol or diesel additional parameters, such as octane number are specified. The boiling point ranges of the various fractions of crude oil coming out of a distillation column and the names of these fractions are depicted in Figure 1.1. The names of fractions, typical number of carbon atoms in molecules and boiling point ranges are also indicated.

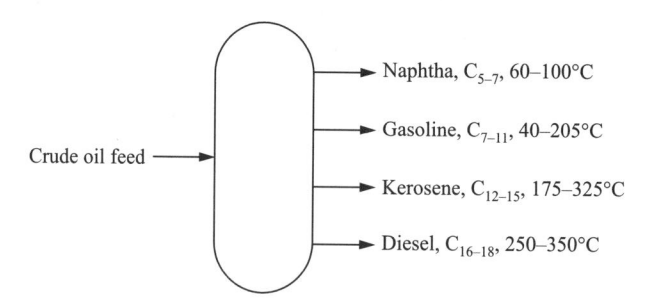

Figure 1.1 Schematic diagram showing the different fractions coming out of a distillation column in a refinery.

In Figure 1.1, the crude oil is separated into different fractions such that each fraction is a mixture of compounds whose boiling points lie in a fixed range. The lightest fraction of the products is LPG, consisting of propane and butane. This is not shown in the figure. The fraction with the next highest boiling point is naphtha. It has the lowest boiling points and is shown coming out at the top of the column. The middle distillates consisting of the gasoline, kerosene and diesel come next. The heavy distillates with the highest boiling points come last. In most refineries the heavy distillates are processed further to give more valuable low boiling point fractions using cracking which we will discuss later.

One of the challenges in the refinery industry is the fact that the nature of the feed, i.e. the crude oil varies from reservoir to reservoir. In fact, the quality of the crude oil coming from a reservoir itself may vary from day to day. The refiner then has to change the conditions of the operation to take into account these variations and obtain an optimum efficiency. To do this he has to have a good understanding of the way operating conditions (temperature and pressure) affect the plant performance since the profit margin in a refinery is very sensitive to these variables. These parameters determine the amount of each boiling point fraction produced.

Another aspect confronting a refiner is that there can be frequent changes in the requirements or demands of the society. This is usually determined by the government policy which varies from time to time. For instance, the requirements may change from an emphasis in production of LPG at one time to diesel production at another time. The refinery must be able to make changes in the operating conditions to meet these fluctuating demands of the consumer keeping in mind that the raw material quality can also fluctuate. All this needs a thorough understanding of the interplay between various conditions in a refinery which affect the separation process.

To meet the demands of the market, additional processes are usually present in a refinery. For example, the fractions in the high boiling point range (typically compounds with a larger molecular weight and size) have a low commercial value. Their commercial value can be increased dramatically if they can be converted into compounds which are smaller in size, lower in molecular weight and which have a lower boiling point. This is accomplished by a process called *cracking*, wherein higher molecular weight compounds are broken (cracked) into lower molecular weight compounds. This can be done using heat (thermal cracking) or in the presence of hydrogen (hydro cracking) or in the presence of a catalyst (catalytic cracking). The exact route used, thermal or hydro or catalytic cracking determines the quality of the products, i.e. whether the compounds with the lower molecular weights are unsaturated, or cyclic, etc.

In the process of catalytic cracking coke (carbon particles) is deposited in the pores of the catalyst. This results in a rapid loss of catalyst activity. Catalysed reactions are conducted in classical packed bed reactors. Here the solid catalysts are stationary and the gas flows around them. These catalysts have a size of around 5–10 mm. Since the objective is to have a high surface area per unit volume one option is to have small particles. This, however, results in a significant pressure drop over the bed. Another approach to increase the surface area over which the reaction occurs is to make the particles porous

10 Introduction to Chemical Engineering

and deposit the active catalyst on the pore walls. This gives a surface area of around 100–1000 m²/g. This approach helps us get a high surface area without any pressure drop limitations. The fluid flows around these catalyst particles and the reactants enter the pores of these catalysts and the reaction occurs in the pores.

When cracking is conducted in reactors which are packed with catalysts the reactor becomes ineffective after a very short time of operation due to coke deposition and this forces the shutdown of the plant. Frequent shutdowns due to coke formation interfere with the continuous operation of other units and this is clearly not desirable.

One of the challenges facing the chemical engineer is the development of catalysts for these cracking reactions and to come up with modifications in the design of the reactors. To overcome the problem of coke deposition during catalytic cracking, chemical engineers have innovated the design of the reactor. Catalytic cracking is now carried out in a fluidized bed reactor. A fluidized state is one in which the solid particles are in a state of dynamic suspension. This modification in the reactor is necessitated to address the rapid fouling of the catalyst by coke deposition. The deactivated catalyst is sent to the regenerator section or unit where the coke is burnt in oxygen. The reactivated catalyst is then recycled back to the cracking unit. The chemical engineer has to determine the flowrates of the gases, the size of the vessel, the temperatures which must prevail and the loading of the catalyst to ensure continuous viable operation. A schematic diagram of the fluidized catalytic cracker is shown in Figure 1.2. In this figure, the flow of different streams from one unit to the other is depicted. The composition of different streams is also shown. Such a diagram is called a *flowsheet*. The regenerator can be viewed as lungs of human body. Here the catalyst is purified with oxygen, whereas in the lungs of the human body blood is oxygenated. The catalyst thus reactivated is sent back to the cracker ensuring a continuous operation.

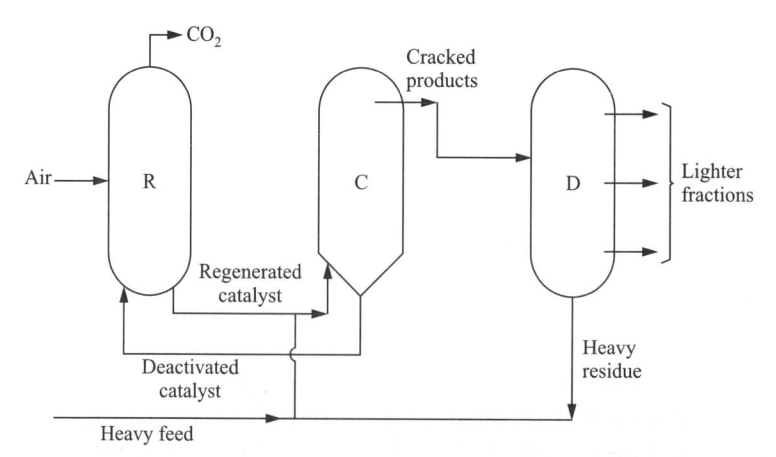

Figure 1.2 A schematic diagram of fluidized catalytic cracker C-Cracker, R-Catalyst regenerator, D-Distillation column.

Crude oil containing more than 0.5% of sulphur is called *sour crude*. In treating this kind of crude oil, sulphur containing compounds have to be removed from the crude oil or its various fractions in a hydrotreater. This is necessary to reduce SO_2 levels in the exhausts of automobiles. To ensure this there are strict norms as to the allowable sulphur content in petrol or diesel which can be used in vehicles. Besides this, it is also important that the sulphur containing compounds be removed since they can act as a poison to the catalysts being used in different processes in the refinery. The sulphur removal is achieved in a hydro-desulphurization unit and is an integral part of the refinery.

Some of the other units present in modern refineries and their functions are as follows:

1. **Catalytic reforming unit:** This unit is used to convert the molecules in the lower boiling point range of naphtha into higher octane reformate (reformer product). The product or reformate has higher content of aromatics and cyclic hydrocarbons. An important by-product of a reformer is hydrogen released during the catalysed reaction. This hydrogen is used either in the hydrotreaters or the hydrocracker.

2. **Fluid catalytic cracking unit:** Catalytic cracking as already mentioned is a process which breaks down the larger, heavier, and more complex hydrocarbon molecules into simpler and lighter molecules by the action of heat and aided by the presence of a catalyst but without the addition of hydrogen. In this way, heavy oils (fuel oil components) can be converted into lighter and more valuable products (notably LPG, gasoline, and other middle distillate components). The catalytic cracking unit is known as the fluidized catalytic cracking (FCC) and is depicted in Figure 1.2. Here the coke deposited in the catalyst pores is burnt off in the regenerator which is a separate unit. Reactivated catalyst particles are recycled back to the cracker through the riser unit. The spent catalyst is drawn from the cracker and is sent to the regenerator. The chemical engineer is interested in designing such a reactor and determining their operating conditions. For this he must understand how the reactants move into the catalyst, and how a continuous flow between the different sections of the cracker can be maintained.

3. **Delayed coker unit:** Delayed coking is a high severity "bottom of the barrel processing" scheme. Here heavy crude oil fractions are thermally decomposed at elevated temperatures to produce a mixture of lighter oils and petroleum coke. The light oils can be processed further in other refinery units or blended into products. The coke can be used as a fuel or in metallurgical applications such as the manufacturing of steel or aluminum.

The Language of the Refiner

Each industry is characterized by its own lingo. The lingo includes the terms which the practitioner uses on a daily basis. An important parameter which is used to describe crude oil is its *density* or *viscosity*. The heavy crude oil has a relatively high density or specific gravity, and a higher viscosity. It is composed of compounds with a higher molecular weight. One of the parameters used to describe the density of crude oil is API gravity and this is related to specific gravity as:

$$\text{API gravity} = \left(\frac{141.5}{\text{Specify gravity}}\right) - 131.5$$

Thus, petroleum crude which has a specific gravity of 1 has an API gravity of 10. The more the specific gravity the lower is the API gravity. Extra heavy crude oil can have a viscosity of up to 10,000 centipoise (compare this value with 1 centipoise the viscosity of water). Hence, this does not flow very easily in a pipe and diluents have to be added at frequent intervals to facilitate the flow.

The petroleum sector discussed so far is one in which chemical engineers play a dominant and vital role. The refineries have a high turnover (in millions of dollars). This is due to the large volume of production. However, this does not imply that they have high profit margins. The profit margins on a per unit production basis are very low and the refiner has to operate his plant optimally so that he can be competitive. He must take into account the fluctuations in the crude quality as well as the demands of the market. Table 1.1 contains the list of major refineries in India, their locations, and installed capacities.

Table 1.1 Refineries in India and their installed capacities

Location	Company	Capacity	Capacity
Jamnagar Refinery	Reliance Industries Ltd.	650,000 bbl/d	103,000 m^3/d
Reliance Petroleum	Reliance Industries Ltd.	580,000 bbl/d	92,000 m^3/d
Mangalore Refinery	MRPL	199,000 bbl/d	31,600 m^3/d
Digboi Refinery	Assam (IOC)	13,000 bbl/d	2,100 m^3/d
Guwahati Refinery	Assam (IOC)	20,000 bbl/d	3,200 m^3/d
Bongaigaon Refinery Assam	IOC	48,000 bbl/d	7,600 m^3/d
Numaligarh Refinery Limited Assam	NRL	58,000 bbl/d	9,200 m^3/d
Haldia Refinery	IOC	116,000 bbl/d	18,400 m^3/d
Panipat Refinery	IOC	240,000 bbl/d	38,000 m^3/d
Gujarat Refinery	IOC	170,000 bbl/d	27,000 m^3/d
Barauni Refinery	IOC	116,000 bbl/d	18,400 m^3/d
Mathura Refinery	IOC	156,000 bbl/d	24,800 m^3/d

Contd.

Location	Company	Capacity	Capacity
Chennai Refinery	IOC	185,000 bbl/d	29,400 m^3/d
Mumbai Refinery	HPCL	107,000 bbl/d	17,000 m^3/d
Visakhapatnam Refinery	HPCL	150,000 bbl/d	24,000 m^3/d
Mumbai Refinery Mahaul	BPCL	135,000 bbl/d	21,500 m^3/d
Nagapattnam Refinery	CPCL	20,000 bbl/d	3,200 m^3/d
Kochi Refinery	Kochi Refineries Ltd	172,000 bbl/d	27,300 m^3/d
Tatipaka Refinery	ONGC	1,600 bbl/d	250 m^3/d
Essar Refinery	Essar-vadinar Gujarat	360,000bbl/d	57,600 m^3/d

The energy requirements of India today are primarily met from crude oil and coal which are obtained indigenously as well as from imports. The reserves of crude petroleum in the world are finite and diminishing quickly. This coupled with the ever-increasing demands of energy as determined by the growing population and the increased standards of living globally has given rise to a strong push for finding alternate sources of energy. This turns out to be an important challenge for chemical engineers since the reserves of petroleum are exhaustible. In this context the prospects of using coal more effectively to meet the increasing demands are discussed next.

Coal Gasification

India has abundant reserves of coal. Several power plants in India are based on coal. Here coal is combusted directly and the heat released is used to generate steam which drives turbines to get electricity. The carbon in the coal leaves as carbondioxide. The efficiency of these coal-based power plants varies from 33% to 48%. This limit comes from thermodynamics and cannot be easily increased. However, the coal found in India has a high ash content and is hence not a clean source of energy. The emissions from power plants in addition to CO_2 also contain particulate matter which is a health hazard. Hence, there is a drive to get a clean source of energy from coal. One possible route to convert coal to a clean energy source is coal gasification. In this, coal is used to produce synthesis gas (a mixture of CO and H_2), which is a good source of clean energy. The Fischer–Tropsch process can be used to produce a mixture of hydro carbons (synthetic liquid fuel) from this synthesis gas.

Synthesis gas is a mixture of CO and H_2. Production of synthesis gas is achieved by gasifying any fuel feedstock containing carbon, hydrogen, and oxygen like coal, biomass, etc. The gasification step is important since it gives us a clean source of energy. By clean we mean the impact on the environment is minimal. For instance, as mentioned earlier the emissions of particulate matter and carbondioxide are high in coal combustion. Should coal be directly used as a fuel, then the impact on the environment would be significant, the particulate emissions would be high. When synthesis gas is used the particulate emissions are drastically reduced.

The gases coming out of the coal gasification process are treated so that they are devoid of any particulate matter. The reactions in the gasifier are carried out in an oxygen-lean environment to ensure that combustion does not proceed to completion. The combustion reactions are exothermic, and give rise to CO_2 and heat. Gasification is normally done in an atmosphere of a gasifying agent such as steam or carbondioxide. The heat released by combustion of the exothermic reaction is used to drive the endothermic reactions of gasification which produces CO and H_2. The products of gasification are essentially CO and H_2. In a typical gasifier unit several reactions occur simultaneously. The two important reactions are as follows.

1. Exothermic combustion reaction

$$C + O_2 \rightarrow CO_2 \quad \Delta_R H = -392 \text{ kJ/mol}$$

2. Endothermic gasification reaction

$$C + H_2O \rightarrow CO + H_2 \quad \Delta_R H = +132 \text{ kJ/mol}$$

The above reactions are heterogeneous and involve the reaction between a gas phase and a solid phase. In addition to this, an important reaction which occurs in the gasifier is the homogeneous (gas phase) reaction, the water gas shift reaction, which provides an additional source of hydrogen, $H_2O + CO \leftrightarrow H_2 + CO_2$.

The mixture of synthesis gas coming out of the gasifier can be used directly in an integrated gasification combined cycle unit to generate power. Here the gases are combusted and the gaseous products are used to drive a gas turbine to generate power. Simultaneously, the heat released in the turbine is used to generate steam which in turn drives a steam turbine generating additional power. Alternatively, it can be processed to give different liquid hydrocarbons. The typical reaction for this is the Fischer–Tropsch synthesis reaction. This can be represented in a general way as follows.

$$(2n + 1)H_2 + nCO \rightarrow C_nH_{(2n+2)} + nH_2O$$

While several kinds of reactors are available for gasifying coal, most of the technologies developed are for coal found in Western countries, which are typically low in ash content. Indian coal has high ash content and hence the technologies developed in the West cannot be directly adopted. This is an example of a situation where we have to be careful so that we do not blindly adopt a technology which is available and find out later that it is not effective.

How can this issue be addressed? One way, this problem can possibly be circumvented is by gasifying a mixture of high ash Indian coal and another fuel such as petcoke or biomass. This will effectively reduce the ash content. However, the two components being gasified will have different reactivities and this has to be addressed in the reactor or gasifier design. The gasification of a mixture of feed stocks is now an area which is attracting a lot of attention.

Since there are a multitude of reactions occurring simultaneously in the gasifier, an important challenge lies in determining the amount of oxygen and steam that have to be used to optimize the performance. One variable which can be optimized is the heating value of the synthesis gas coming out of the reactor. One cannot use stoichiometry arguments since several reactions occur in parallel and the extent to which they proceed depend on each other, i.e. the reactions are coupled. Both oxygen and steam are expensive reactants, i.e. oxygen has to be obtained by separating it from nitrogen in the air using cryogenic distillation and steam needs energy for its production. Using a high amount of oxygen would result in a lower production of synthesis gas. Here a lot of coal will be combusted directly to CO_2 and not enough coal will be left for gasification. Similarly, using a low amount of oxygen would result in a low amount of heat being liberated and, consequently, the heat supplied for the endothermic gasification reaction would be low and insufficient. This again would result in a lower amount of synthesis gas being produced and leave some of the coal unreacted. The calculation of the optimal amount of oxygen and steam required for a fixed amount of fuel of known composition is one of the issues in gasifier operation.

Euro Norms/Bharat Stage Norms to Curb Atmospheric Pollution

Health care is high on the agenda of any government. The quality of the air we breathe is extremely important since it has a direct effect on our health. The adverse effects of the environment on the population can be felt in a very short time span. The number of vehicles on the streets has increased and with this their contributions to the deteriorating quality of air has gone up. The Euro norms have been proposed to regulate the emissions of automobiles. These norms also have implications on the performance of refineries since the exhaust emissions from vehicles also depend on the quality of fuel used. Hence, specifying the exhaust emissions has implications on engine design as well as the quality of the fuel. It may not be possible to meet the norms with a particular quality of fuel no matter how well the engine is designed. For instance, if the fuel has a high sulphur content the exhaust emissions will have high SO_x emissions. It is, hence, necessary that the sulphur containing compounds are eliminated in the refinery, using a hydrotreating process like hydro-desulphurization. Similarly, if the engine design is not good, no matter how good the fuel is, emission norms may not be met since the emissions may now contain unburnt hydrocarbons. The onus for reducing these emissions lies with both the refinery industry which controls the quality of the fuel and the automobile industry responsible for the engine design. The interplay of the roles of the chemical engineer and the mechanical engineer in controlling air quality is shown in Figure 1.3.

16 *Introduction to Chemical Engineering*

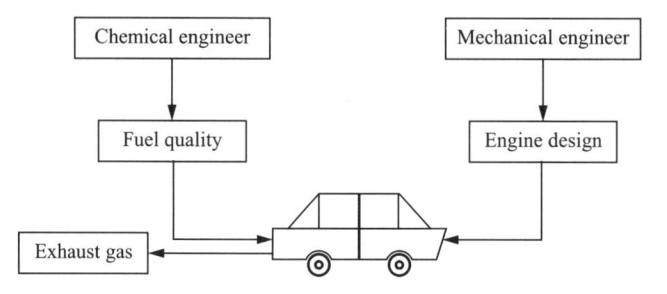

Figure 1.3 Role of chemical and mechanical engineers in determining air quality.

The growth in the economy of India has resulted in an increase in the buying power of the common man. As a result, the number of cars and motorized vehicles on the road has increased significantly. It is, hence, necessary that measures to control pollution levels are formulated and implemented effectively. For this a clear understanding of the roles of the different contributors is required. Emission norms from vehicles specify the amount of carbonmonoxide CO, hydrocarbons (HC) and nitrous oxides NO_x that can be released into the atmosphere. Typical norms are shown in Table 1.2. The Euro norms are available for vehicles running on petrol and diesel. The standards in India are different from those of Europe. This difference is necessary since it takes into account the variations in the quality of fuel available in India and the testing methodology. The corresponding or equivalent norms in India are called Bharat Stage (BS) norms. To explain why this is necessary, consider as a specific example Euro III norms. Here the testing is done under sub-zero temperature conditions while the average annual temperature in India is 24–28°C. Carrying out the testing under sub-zero conditions is not relevant under Indian conditions. Another point is that the maximum speed in testing for BS III is 90 km/h while that in Euro III is 120 km/h. This is necessary as the road and traffic conditions which prevail in India are such that the average speed is lower than that in the West. Consequently, the norms we follow in India are the BS norms which are different from the Euro norms. Currently in India BS IV norms have been implemented in select cities (the metros) since April 2010. The automobile manufacturing companies and refineries have to be geared up for meeting these strict emission norms. Table 1.2 shows the norms of vehicular emissions from Euro I and Euro II.

Table 1.2 Typical norms for vehicular exhausts

Species	1998	Euro I	Euro II
CO (carbon monoxide)(gm/km)	4.34	2.75	2.20
HC + NO_x (g/km) Hydrocarbons and Nitrous oxides	1.50	0.97	0.5

As discussed earlier the responsibility of meeting the emissions levels from vehicles is determined by the chemical engineer (quality of the fuel) and the mechanical engineer (engine design). The auto-fuel policy of the Government of India specifies the characteristics of fuel (metal, nitrogen and sulphur content) in quantitative terms. One of the key parameters specified in the auto-fuel policy is the sulphur content in petrol or diesel fraction. Typical values of this are given in Table 1.3. Heavy crude oils have to be processed in the refinery to remove metals, nitrogenous and sulphur containing compounds. These compounds can act as poisons to catalysts in downstream operations in a refinery and, hence, it is important to remove them. This also results in lower emissions of toxic gases from exhausts.

Table 1.3 The norms for sulphur content in gasoline/diesel

	BS II	Euro III	Euro IV
Petrol	500 ppm	150 ppm	50 ppm
Diesel	500 ppm	350 ppm	50 ppm

To meet the tight control in the sulphur content in the autofuel as specified in Table 1.3, is the responsibility of the refinery. The sulphur compounds present in the feed stock are removed by converting them to hydrogen sulphide through an operation called *hydrodesulphurization*. Here hydrogen is passed along with the fuel cocurrently on a catalyst bed of cobalt and molybdenum. The sulphur containing compounds react with the hydrogen and produce hydrogen sulphide. The sulphur containing compounds are a diverse mixture of various hydrocarbons. During the hydrodesulphurization process all these compounds have to lose their sulphur. A few of the compounds get desulphurized easily. The compounds which can be converted easily are responsible for the initial reduction of sulphur content which is rapid. These include mercaptans and quinolines. As we seek further removal of sulphur content, it becomes more challenging. Consequently, the reduction of sulphur content in diesel or petrol to Euro III norms is achieved much more easily as compared to meeting Euro IV specifications. To meet the latter specifications the more difficult compounds (the ones which are less reactive) have to be converted to sulphur free form. The examples of such compounds are dibenzothiophenes. This may call for more stringent conditions of temperature and pressure in the hydrodesulphurizer. It may also demand a new catalyst to be developed which can withstand these stringent conditions.

Both fuel quality and design of the engine determine the emissions from a vehicle. The question that confronts us is: is it better to invest money in a refinery to get a good quality of fuel or should the investment be made to get an improved engine design in the automobile sector? In other words, is it preferable to invest money and develop technology by way of new catalysts in the refinery sector to remove the sulphur bearing compounds, etc. or in the automobile sector to ensure that the products of combustion are not going to harm the environment? For instance, in India, all the refineries have desulphurization units to remove sulphur. At the moment SO_x levels in the atmosphere are low and pollution due to them is not a problem. Should SO_x emissions be an issue, we may have to go in for ultra low sulphur content in the diesel or petrol fractions. However, NO_x levels are high in the atmosphere and the primary contributors to this are vehicular emissions. To reduce NO_x levels we can do one of two things. First, we can add units in the refineries to remove nitrogen containing compounds, i.e. carry out a hydrodenitrification. Another possibility to reduce NO_x levels is to modify the design of exhaust treatment devices (the three-way catalytic convertor TWC) in vehicles so that the pollutants (NO_x) are converted to harmless compounds (N_2) and then released into the atmosphere. This involves making investments in the automobile sector and addressing the tuning of carburettors, changing the design of engines, designing trimetallic coating of catalysts in the exhaust gas convertors, etc. The solution to this problem is not unique as the final decision is based not only on technical issues but also on socio-economic factors.

At present the removal of sulphur containing compounds is done primarily at the refinery stage. As a result of this the SO_x levels are very low in the atmosphere. The removal of nitrogen bearing compounds is done in the catalytic convertors which treat the exhaust gases coming out of automobiles. The NO_x levels are relatively higher in the atmosphere in India as compared to SO_x.

Versatility of a Chemical Engineer

The common man has a view that the chemical engineer has a role to play only in big factories and refineries. Now we take two examples, which show the versatility of the chemical engineer. These examples are from areas which you may only remotely associate with chemical engineering. You will see two areas or industries in which the chemical engineer plays an important role which you may not have dreamt of at all. These examples are chosen to show how the interdisciplinary nature of chemical engineering allows a graduate to fit into almost any industry.

Semiconductor processing

In the electronics industry the first microchip or integrated circuit (IC) was created in the second half of the 1950s. Today these chips are getting smaller day by day. In addition to this, it is also possible to pack more information in them compared to the larger chips in the earlier days. Chemical engineers have contributed to the development of these chips. They are involved in developing new materials and coming up with suitable processes for chip manufacturing.

Semiconductors are materials whose ability to conduct electricity lies between that of a conductor and an insulator. Germanium was the first known semiconductor. The present-day industry exploits the semiconducting properties of silicon. First a pure mono crystalline form of silicon is produced in the form of an ingot whose diameter lies between 6 and 12 inches. The wafers of thickness around 600 microns are sliced from this ingot. A slurry, i.e. suspension of nano-sized particles is used to polish these wafers. This is then subjected to several steps where different layers of materials are added on to the surface. The different materials added could be semiconductors, insulators, and conductors. These layers make up the capacitors, resistors, and transistors which constitute an integrated circuit. The circuits, thus, made are what we use in our cell phones, computers, and iPods.

As the chips become smaller with more transistors packed in a chip, the demand for the level of purity of the materials used in the processing has increased. These demands are even more stringent than what is used in the food processing industry. Chemical engineers have to come up with methods which help in maintaining high levels of purity. The materials used for the construction of equipment, vessels, pipes where the manufacturing is done have to be clearly specified and the process has to be closely monitored.

After the integrated circuit is identified and designed, the focus shifts to manufacturing them on a commercial scale. Here chemical engineering principles like chemical kinetics, fluid mechanics and transport phenomena are used. More than a few hundred steps are involved in creating a chip from the silicon wafer. Of these, chemical engineers play a major role in developing and implementing deposition (chemical vapour deposition, electrochemical deposition, spin-on coating), removal of excess materials (wet etching, dry etching or plasma etching, chemical mechanical planarization), cleaning (removal of contaminants from the wafer using hydrofluoric acid, ammonia, hydrogen peroxide) and material modification (oxidation of silicon to silicon dioxide using oxygen or steam, diffusion of boron or phosphorous in silicon to form different types of semiconductors).

In chemical vapour deposition (CVD), the reaction is carried out using vapour phase reactants while the desired product is formed on the wafer, in the solid phase. Depending on the type of material to be deposited, the formation may be controlled by mass transfer or by reaction. CVD processes are operated either at high pressure and temperature (atmpospheric pressure CVD or APCVD) or at relatively low pressure and temperature (low pressure CVD or LPCVD). In APCVD, the reaction is fast and mass transfer is slow, whereas in LPCVD reaction is slow. In certain cases, the presence of plasma is required to carry out the reaction and plasma enhanced CVD (PECVD) is used. The use of metallo-ogranic reactants for forming metal films is called the *metallo-organic CVD* (MOCVD). Different techniques are used to obtain films of controlled thickness and desired properties.

Copper wires are used to connect the transistors. Cu is deposited in very thin wires, with thickness of the order of 100 nm, using electrochemical deposition. In order to ensure proper deposition, the composition of the electrolyte and the applied voltage are carefully monitored and controlled. For creating insulating materials around the copper wires, organic films are used. They cannot be deposited using high temperature techniques since the high temperature results in degradation. Hence, they are dissolved in a solvent which is poured on the wafer. A thin layer of the solution is formed on the wafer by rotating the wafer with the solution on the top. The solvent evaporates leaving the organic coating. This technique is called *spin-on coating*.

To remove the excess materials, a suitable chemical is used. For example, to remove silicon dioxide, HF is used, whereas to remove silicon nitride without affecting the silicon dioxide, hot phosphoric acid is used. Wet chemical etching offers very good control over the choice of material to be removed, but it is difficult to control the thickness of material removed. Many times, anisotropic material removal is necessary. For example, a column of material may need to be removed to create a metallic connection. In those cases, wet etching is not suitable. Instead, reactants in gas phase, along with plasma created by a high voltage are used. This is called *dry etching*. In dry etching, the normal gas phase component does not react with the solid. However, in the presence of a high voltage, a plasma which contains highly reactive radicals is formed. The radicals react with the solid and form gaseous products. It is easy to control the etching here, since the plasma can be switched off easily. With suitable modifications, dry etching can be used to obtain anisotropic etching. Wet etching is also used to clean the wafer surface. Dust particles may fall on the wafer and if they are not removed, will cause problems in the subsequent steps. Treating the wafer with ammonia and hydrogen peroxide will remove the inorganic contaminants. Acid and hydrogen peroxide will remove the metallic contaminants. Etching the surface with hydrofluoric acid will remove some silicon dioxide and also dislodge particles present on the surface. Thus, cleaning can be thought of as a special case of wet etching.

In certain processes, it is necessary to remove excess material and also obtain a very planar surface. In those cases, a mixture of abrasives and chemical is used to polish the wafer. Here, the chemicals will not usually etch the material; however, they will modify the surface and form a soft layer on the top. The mechanical abrasion will cause the soft layer to be removed. Thus, a chemical action or the mechanical action alone is not sufficient to remove the material and form a planar surface, but the combined action causes the removal of excess material and results in a planar surface.

The silicon material is converted into different types of semiconductors by adding small quantities of materials called *dopants*. If boron is added to Si, the silicon is called P-type silicon and if phosphorous is added, it is called N-type silicon. The quantity and the depth of doping must be carefully controlled. While the initial addition is done by a process called the *ion implantation*, further movement of the dopants in the solid, to obtain the final controlled doping, is done by heating the wafers to high temperatures for a controlled amount of time. The dopants will diffuse in the solid, and the diffusion is calculated using the basic laws of diffusion used in chemical engineering. High temperatures are used to obtain reasonable diffusivity values since solid-solid diffusivity values at room temperatures are very low.

In some steps, silicon is converted into silicon dioxide. If the silicon dioxide is used in the transistor, then a high quality oxide is needed and dry oxygen (without moisture) is used. If the silicon dioxide is going to be only used as an intermediate, then steam is used. Dry oxidation is slow but can give a good quality oxide, whereas wet oxidation is fast but will give a porous oxide. The oxidation process is limited by the rate of reaction in the beginning and by mass transfer at later stages. The control of the entire operation relies on some of the basic principles of chemical reaction engineering (i.e. gas diffusion in solid, followed by reaction).

The entire processing of these chips is done in clean rooms. These rooms have strict specifications such as one particle in a cubic foot (compare this with the requirement in a hospital of 10,000 particles in a cubic foot). If a dust particle is lodged in a circuit, then it would completely damage it since it would obstruct the pathway of current. In other words, in the small scale that we are talking about, this tiny spec looks like a huge football. Chemical engineers design the clean rooms which are specified by the number of particles per cubic foot. Filtration systems are designed to capture chemical vapours, microbes, and dust particles. These fabrication facilities are called *fabs*. Here pressurized air is used to make sure that the tiniest of particles, if suspended on the chip, can be carried away. The stringent criteria for cleanliness of the room are to ensure that the chip does not get contaminated. Figures 1.4 and 1.5 show the pictures of the clean room facility available in IIT Madras.

Figure 1.4 Photograph of Class 10,000 clean room in IIT Madras Particle Science Laboratory.

Figure 1.5 Air shower connecting Class 10,000 and Class 100 cleanrooms.

Role of Chemical Engineers in Biomedical Engineering

Several areas of biomedical engineering such as haemodialysis have advanced due to chemical engineering. This is one of the treatment techniques used on patients suffering from renal failure (when the two kidneys fail to function). Failure of the kidneys results in the accumulation of toxins like creatinine, urea, and water in the blood stream. It is, hence, necessary to remove these from the blood stream through some artificial means. The principle on which dialysis is based is that of transport of species across a semi-permeable membrane. The two streams, blood and dialysate (the fluid used to extract the toxins) flow on either side of a membrane. The dialysate consists of a solution of several salts such as chlorides of sodium, magnesium, potassium and calcium, and sodium bicarbonate. The flow is counter current to increase the process (the rate of solute transfer) efficiency. Here the toxins move from the blood to the aqueous phase when the pressure is reduced in the dialysate compartment.

The dialyser is typically a rigid cylinder which houses fibres made of a proprietary polymer. The area of the fibres available and the flow rates of the various streams are optimized for efficient solute transfer or extraction. A schematic diagram representing the flow of the various streams and the transport of ions is shown in Figure 1.6. The dialysis unit

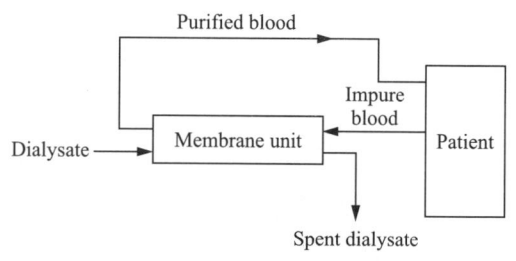

Figure 1.6 Schematic diagram showing haemodialysis. Blood and dialysate flow in a counter current inside the membrane unit.

can be a portable one. The patient is connected to the dialyser till the concentrations of toxins in the blood reduces to acceptable limits. He can then be taken off the dialyser. He is connected to the dialyser again when the concentrations of the toxins in the blood reaches a high value.

Chemical engineers can predict the rate at which the toxins are transported across the membrane for a given combination of flowrates and dialysate composition and membrane characteristics. This helps them design these units efficiently. This example shows how chemical engineers can help in developing life-saving applications.

Similarities in Dissimilar Applications

In this chapter we have seen various industries, situations, and applications in which a chemical engineer plays a vital role. The chemical engineer focuses on the underlying principles on which the various processes function. This helps him understand the conceptual similarities lying in apparently dissimilar processes.

The example of dialysis shows how membranes can be used to purify blood. Solutes are transported selectively across these membranes here. The principles of ultra-filtration, nano-filtration, and reverse osmosis used in treating wastewater coming out from a chemical industry are also membrane-based processes. These processes and dialysis used to purify blood are similar. It is the understanding of the basic mechanism and principles of a process which help a chemical engineer find novel solutions to technological problems. He is able to use ideas from one application and extend them in finding solutions in a different area. We now discuss different membrane separation processes used in industries.

In any membrane separation process, there are two outlet streams: the retentate and the permeate. The *permeate* is the stream which flows or permeates through the membrane and the *retentate* is the stream which contains material which cannot flow through the membrane and is retained by it. Separation and purification processes based on membranes can be broadly classified into three different categories.

Reverse osmosis

Reverse osmosis, also known as RO, is a common technique used for desalination, i.e. removing salt from sea water and making it potable water, safer for human consumption.

Consider a membrane which separates two solutions of a solute with different concentrations. *Osmosis* is the process in which the solvent moves from the region where the solute concentration is low to the region where the solute concentration is high. The motion is such that the system tends to evolve into a state where the concentrations are same in both the compartments or the concentration difference in the two compartments is reduced. This is dictated by thermodynamics and here the equilibrium state is one in which no further transfer of solute can occur. The reduction in concentration difference can happen only through the motion of solvent since the solutes cannot pass through the membrane. The solvent moves from the dilute solution side to the concentrated solution side until the equilibrium is reached. This is called *osmosis*. This process cannot be used to purify water since the more concentrated solution gets diluted.

Purification of water can be obtained using *reverse osmosis* (RO). Here the direction of the solvent flow is reversed by applying a pressure greater than the osmotic pressure. This is achieved by increasing the pressure on the high solute concentration side. In a desalination application of reverse osmosis the concentration of salt in salt water increases beyond the feed value while we get pure water on the other side. The pressure required for desalination of sea water is around 40–70 bar and the membranes must be able to withstand this high pressure.

In reverse osmosis, we separate large molecules and ions from a solution by applying a high pressure on the saline solution which is on one side of the membrane. The membrane has pores which do not allow the larger solute molecules to go through but can allow the smaller solvent molecules like water to go through.

Nano-filtration

Nano-filtration operates on the same principle as reverse osmosis. The key difference is the degree of removal of monovalent ions such as chlorides and calcium. In reverse osmosis we have almost a complete removal of the monovalent ions. In nano-filtration, the removal of monovalent ions is not complete. Some monovalent ions can be found in the permeate. This is determined by the choice of the material of the membrane. In nano-filtration the permeate has monovalent ions, whereas the retentate has all solutes but monovalent ions. This operates at a slightly lower pressure of 10–40 bar. There are a variety of nano-filtration membranes available with each type being particularly suited to a specific application. Nano-filtration membranes and systems are used for water softening, food, and pharmaceutical applications. In the dairy industry, the nano-filtration process is used to concentrate and partially demineralize liquid whey.

Ultra-filtration

Ultra-filtration is a separation process using membranes with pore sizes in the range of 0.1 to 0.001 micron. Here the permeate has minerals and the retentate has proteins and fats. The operating pressure is around 2–8 bar. Typically, in ultra-filtration high-molecular-weight substances, colloidal materials, and organic and inorganic polymeric molecules are retained on the membrane. Low molecular-weight organics and ions such as sodium, calcium, magnesium chloride and sulphate can permeate the membrane. Because only high-molecular-weight species are removed, the osmotic pressure differential across the membrane surface is negligible. These membranes are specified by molecular-weight cutoff (MWCO). For instance, a membrane that retains dissolved solids with molecular weights of 10,000 and higher has a molecular-weight cutoff of 10,000. The molecular-weight cutoff is a useful guide when selecting a membrane for a particular application. Ultra-filtration is used in pretreatment of water for micro-electronics industry, in dairy industry, in polyvinyl alcohol industry, and also in harvesting enzymes.

We see how two completely diverse examples, one based on desalination of sea water and the other based on the removal of creatinin from human blood are based on the same principle. As chemical engineers we understand the commonality in the principles of operation of these two apparently diverse systems. This knowledge helps us analyse these systems and allow us to come up with new applications in different areas of technology.

Note to the teacher: This chapter discusses only qualitative concepts. One approach is to have the students submit assignments on various topics. Assignments can be submitted individually or by a group of students. The students need not be expected to write verbose reports as part of their assignments. They should be encouraged to think laterally and be told that in several situations there are no unique correct answers to any of the questions. In fact, it is expected that different groups evolve to different solutions to the same problem. The objective is to encourage the group to think about a problem. This will help them focus and synthesize thoughts and write a clear, precise answer. The only way technology can develop or any change can occur is when there is a fresh approach to a problem.

Note to the student: Most of you would have seen movies. At the end of the movie you form an opinion of it being good or bad. Directing a movie is similar to writing a report. The author of the report is like the director of a movie. The sequence of ideas in a report or presentation decides how good it is. Report writing and oral presentation skills need to be developed in students. This comes with practice and effort. Having sound technical knowledge alone is not enough. It is important to develop communication skills and express the knowledge you possess cogently and coherently. The first draft of the report and similarly first draft of the presentation may not be the best. You must be prepared to work on these till you are satisfied with the quality. Tips on writing effectively can be found in Suraishkumar (2004).

Today, a lot of information is available on the Internet on various topics. Once students are asked to write a report or make a presentation it is necessary that they do not indulge in plagiarism. The students should avoid doing a copy and paste from an article on the Web. It is necessary to read and get all the information on the subject topic from the Net and express it in your own words. Make sure you refer to the articles that you have used and if you are going to reproduce something directly from an article take prior permission from the original author or the source.

Exercises

1. Choose one of the products (soap, shampoo, etc.) that you use on a daily basis. Find out the chemical composition, and write the properties that you expect it to have. Submit a one-page report containing the above details.

2. Choose a product from the market. Note down its price. Estimate the cost of the raw materials that go into its preparation.
   (a) Estimate what will be the profit on each unit of the product.
   (b) Do you think this is the right estimate? If not, give the other factors which you have ignored in the profit calculation.
   (c) Now discuss why chemical plants have to be large in size from economic considerations.

3. Consider an outdoor swimming pool. The entire floor and walls of the swimming pool have tiles. To maintain the level of water in the pool everyday some water has to be added.

   What is the reason that this water has to be made up? Is it possible to estimate this make-up? Explain qualitatively how will you do it. How will you verify if your reason is correct?

4. In Problem #3, it is found that there is a sudden increase in the make-up water required above that which you had estimated. Can you hypothesize a likely cause for this? Frequently you come across situations where things behave differently from your expectations. You need to understand why this happens. So you make a conjecture or hypothesis and then see if you can verify it.

5. In August 2011, the ship MV Rak sunk in the Arabian Sea around 20 miles off the coast of Mumbai. Write a report on the causes of this event, and the damage to the environment caused by it.

6. What is the motivation behind carrying out cracking in a refinery? Explain the difficulty in catalytic cracking and its disadvantage. Discuss how it is overcome.

7. To combat the problem of atmospheric pollution discussed, two possible strategies can be adopted. What are these? Discuss their pros and cons.

8. Explain how does haemodialysis work?

9. Why do we follow BS norms in India and not Euro norms?

2

Modern Chemical Engineering Plants

Introduction

Chemical engineering is concerned with the manufacture of chemicals/pharmaceuticals and the understanding of all physical and chemical steps that constitute the processes involved in it. A firm understanding of the principles of the process is required so that improvement in technologies to render it energy efficient, cost effective and environmentally friendly can be achieved using a holistic approach. The engineer is interested not only in understanding the qualitative behaviour of a process but also in predicting the performance quantitatively. He would be interested in knowing the maximum temperature rise in a reactor sustaining an exothermic reaction, for instance. Alternatively, he may want to determine how much of the reactants fed into a reactor get converted to products. He can measure the performance experimentally and compare it with predictions based on an ideal behaviour. This would help him understand how close the real system is to the ideal or desired performance levels. The quantitative prediction of the performance of reactors and other units is beyond the scope of this chapter.

In this chapter the broad features or characteristics of modern-day chemical process plants are discussed. This helps us obtain an overall perspective of factors governing a chemical process plant. *Batch processing* as well as *continuous processing* of chemicals is first discussed. Two case studies of the processes are then presented to explain how economic and environmental factors determine the evolution of a specific industry. These will help the student understand that technology involved in the manufacture of a chemical changes dynamically, i.e. with time. The factors which influence this are environmental issues and economic issues.

Batch Processing

In the beginning most chemicals were manufactured in the batch mode. In the batch mode the raw materials are mixed in a vessel and brought to the right conditions of temperature and pressure to initiate the reaction. In the vessel called a *batch reactor* a chemical transformation takes place as a result of which the reaction products are formed. The time taken to form the products depends on the rate of reaction. If the reaction were slow, the time taken for products to be formed would be high and vice versa. The reaction can be accelerated by using a catalyst. The identification of a catalyst is an integral part of reactor development. After sufficient amount of products has been formed the contents of the reactor are emptied. The products are then purified by further downstream processing and transformed into a commodity of commercial interest.

On most occasions, of course, the process is more involved and requires several steps. For instance, after the reaction the contents of the reactor would consist of a mixture of unreacted reactants and the products. It would, hence, be necessary to separate the different species from this mixture. This separation could be a straightforward physical process which exploits property differences like boiling points if the products are liquids or it could be more involved. After the separation they may be subject to further reactions and separations. *Paint manufacture* is done through a batch process which involves no reactions but only mechanical operations.

Let us consider a reaction involving two liquids giving rise to a product which is also in the liquid phase. For simplicity we assume all the three liquids are miscible. This means that there is only one phase in the reactor and all the species or components are in that phase. As an example of such a system consider the liquid phase reaction between sodium hydroxide and hydrogen chloride. Here the reactants are miscible and the products formed are also in the liquid phase, i.e. the system is homogeneous. (An example of a two-phase system or a heterogeneous system is a mixture of kerosene and water. Here there are two phases, the organic phase and the aqueous phase, and they retain their identities.)

Some of the questions which the engineer has to answer in batch processing are:

1. What should be the concentration of the reactants that are fed to the reactor? What happens if the concentrations are too low? What if they are too high?

    If the concentrations are too low, then the reaction will be very slow and the time taken for products to form will be high. If the concentrations are high and the reaction is strongly exothermic the temperature will rise very sharply and this could lead to vaporization of the reactants. This in turn can lead to a pressure build-up in the reactor and a possible explosion/accident.

2. What should be the sequence of adding the reactants and carrying out the different steps in the reactor and how should the provisions for this be made? This is important when two reactants can react and give several products of which one is desirable and the others are not.

3. What should be the volume of the reactor for carrying out a reaction? How much time does it take to get a desired amount of products?

   The first question is determined by the amount of products required and the second by the speed of the reaction.

4. Do we need external heating or cooling for the reactor?

   This is determined by the endothermic or exothermic nature of the reactions along with the need to control the temperature.

5. What should be the material of construction of the reactor?

   This should be chosen such that it does not get corroded or react with reactants or products. The material must be chemically inert. In a laboratory a student was pumping benzene and a mixture of sulphuric acid and nitric acid using two pumps. A nylon adaptor was used to connect the tubing to the pump. After a couple of days of carrying out the experiment the student found that the nylon adaptor had disappeared. His first thought was that somebody might have taken it away without his knowledge. Later on he realized that it had dissolved in the liquid being pumped. This example brings out the importance of material of construction. One can learn by trial and error and trying out different things. Alternatively, one can read about the experiences of others or read and gather relevant scientific knowledge and proceed to design a successful experiment. One way to avoid corrosion problems is to have a lining of an inert material along the interiors of the reactor, e.g. glass lined steel reactors.

Several products even today are manufactured in the batch mode. These include speciality chemicals, pharmaceuticals, paints, etc.

Some important features of batch processing are:

1. **Intermittent operation:** The main feature of batch processing is that the operation is intermittent or discrete. The products are available for withdrawal from the reactor or the unit only at a specific instant. This depends on the time taken for the various processes which occur in that unit in the manufacture. Thus, if the reaction time is eight hours, we will not get any products till the entire step is completed. Products are harvested every eight hours.

2. **Dynamic nature:** Each step of the batch process is a dynamic step. By this we mean that various dependent variables like concentration or temperature vary with time. For instance, in a reactor the concentration of products would increase with time as the reaction progresses. Similarly, if the reaction were to be exothermic the temperature of the reactor would increase with time if the reactor is insulated and there is no temperature control.

3. **Uncertainty in quality:** Operations were performed and monitored manually in the early days. In manual operation the human element plays an important role in the manufacture and this results in variations in the quality of the product from one batch to another. For example, consider a process which requires that the vessel be preheated to a temperature of 400 K at the beginning. If the operator is in a hurry he may not have been patient enough for the temperature to reach 400 K but may have started the process at 375 K. This is likely to adversely affect the quality of the product.

4. **Small scale production:** Batch processing is efficient when the scale of production is low. This helps in better control of processes and quality of products. It can be used for production of "high value" chemicals, i.e. antibiotics, perfumes, etc. Here the volume of production is low and the emphasis is on quality.

5. **Multiproduct manufacture:** Batch processing offers flexibility and scope for optimization by optimally using the resources available. For instance, if scheduling of various operations is planned in advance, then it is possible for several products to be manufactured using the same equipment inventory. This is usually exploited in the pharmaceutical industry.

The better quality control achievable at small scales can be understood by a simple analogy. Cooking for a small family in a household results in a better quality of food. (Remember your mom's cooking is always good.) This is because it is easy to control the quality of the food. Here the contents of the vessel can be well mixed since the cooking pot (which is essentially a reactor) is small and we have a uniform quality (no spatial variations) inside the vessel. Contrast this to the quality of food you have in your hostel mess (or dormitory canteen) where the food is cooked in bulk. Larger cooking pots are used, mixing is poor and you can have spatial variations in the amount of cooking in a vessel. As a result, some portions of the food may taste different from the others. Here it is difficult to control the quality in these systems. Several industries rely on batch processing. For instance, paints, perfumes, and pharmaceuticals are manufactured using batch operations. We now consider a typical batch process system and describe the steps involved in it.

Paint Manufacture

The manufacture of paints and other coatings is done through a series of operations which are carried out in the batch mode. In-between batches, equipment cleaning is necessary. Liquid paints are made of a finely divided pigment dispersed in a liquid composed of a resin or binder and a volatile solvent. Hence, there are three main raw materials: pigment, binder, and solvent. The pigment is responsible for the colour of the paint. Figure 2.1 shows the steps involved in the manufacture of paints. The arrows between the blocks do not indicate the flow streams but the sequence of steps in the batch process.

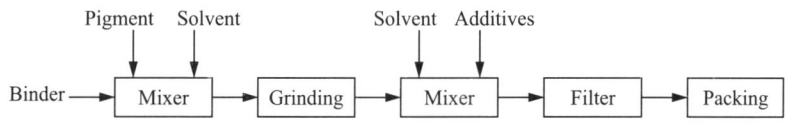

```
            Pigment  Solvent        Solvent  Additives
               ↓       ↓              ↓        ↓
Binder →   [ Mixer ] → [ Grinding ] → [ Mixer ] → [ Filter ] → [ Packing ]
```

Figure 2.1 Steps involved in paint manufacture.

Most of the operations are mechanical and are not based on chemical transformations. Chemical reactions are involved in the manufacture of the constituents of paint but not in the manufacture of paint itself. The various steps in paint manufacture consist of mixing, dispersing, thinning and adjusting properties and filling containers. The processes employed help control the colour and properties of paints. The raw materials are liquids, solids, powders, and slurries.

The first unit operation is mixing. Here vegetable oils such as linseed oil, pigments (titanium dioxide) and fillers like calcium carbonate, and plasticizers are weighed and fed automatically to the mechanical mixers. Here the aim is to get a mixture with a uniform composition.

This is then taken to the next unit for *grinding*, *mixing* and *homogenizing*. In grinding the size reduction of pigments takes place. The mill used for grinding depends on the type of pigments, fillers, etc. Ball mills and steel roller mills are used extensively in grinding. The batch is then transferred to a mixer for *thinning* and *dilution* where solvents and other additives are added. This is used to control the properties such as viscosity of the paint.

After this it is filtered or centrifuged to remove any non-dispersed pigments and entrained solids. Salts of metals such as cobalt, lead, and zirconium are added to enhance the drying. The paint is poured into cans, labeled, and then packed.

The Transition from Batch to Continuous Processing

With the advent of the Industrial Revolution the demand for chemicals increased drastically. There was an increase in the requirements of chemicals which drove the production up. In addition to this there were demands on the quality of chemicals. Better quality implied to higher purity. Thus, instead of being satisfied with sulphuric acid of 68 weight per cent purity there was a demand for 98 weight per cent sulphuric acid. This increase in the demand of chemicals forced a shift in the chemical industry from batch to continuous processing.

From the point of view of economics it was better to use continuous processing for large-scale production of chemicals. Here the fixed costs increased since the capital outlay for the large-scale equipment was higher. However, the operational costs of the process per unit of the product were low. In the long term the costs of running the plant were brought down and this made it a viable option. In a large-scale plant there is a more efficient consumption of utilities like electricity and steam resulting in a reduction in the variable costs. The increase in size made it possible to interlink different streams with a view to recovering energy, etc. For instance, the heat liberated from an exothermic reaction could be recovered and used to generate steam. This would be used to drive a turbine and help in recovering energy (as is done in sulphuric acid plants).

The continuous production of chemicals is similar to that seen in an assembly line. Here chemicals are pumped continuously into each unit of the plant and the products are withdrawn continuously from that unit and sent to the next unit.

The continuous operation of a plant implies that it is running for 24 hours/day, seven days/week. Typically, a continuous plant runs for 335 days of the year. The plant is closed down for 30 days of the year for maintenance to ensure that the plant can run for 335 days of the next year. The plant is also less dependent on manual labour and its control can be done automatically through a centralized control system. Here the performance of various units, as measured by the temperatures and other parameters of various streams and units, is logged into a computer automatically. This helps us detect if there are any faults in the operation sufficiently early so that preventive, anticipatory measures can be taken.

The features of a continuous plant are as follows.

Continuous production: Here the raw materials are fed, i.e. pumped continuously into different units and the product stream is withdrawn continuously from the plant.

Steady state: In this mode of operation there is a brief period during the initial stages (called the *start-up*) and during the final stages, called the *shutdown*, when the system is in a dynamic state, i.e. the different variables at a point change with respect to time. During the start-up the operator starts various pumps and turns on various instruments in a particular sequence or a predetermined protocol. After the start-up period is over the plant reaches a steady state after some time. Here the different dependent variables like pressure, temperature, and velocity at a given point in space in the unit do not change with respect to time. Hence, if there were probes to monitor these variables they would show a constant reading. In a typical chemical plant there is good instrumentation to help monitor the progress of the plant. There are also controllers to ensure that if there is a drift in any of the variables, then the appropriate action is taken. If the temperature in a unit increases beyond the desired value, for instance, a coolant is circulated so that the temperature is brought back to normal. This helps in guaranteeing the quality of the product.

Automatic control: In the continuous mode of operation it is easier to implement automatic or computerized control strategies to negate the effect of disturbances which can result in malfunctioning of the plant or a unit. Consequently, this reduces the dependence on manual labour and the quality of the product is better. For example, if the temperature of a unit is increasing, then the temperature sensor would measure this and a control action to increase the supply of the coolant to reduce the temperature would be automatically initiated. These control actions are usually "easy" since the variable has to be controlled at a fixed value, i.e. the steady state. This is in contrast to the batch process where the variable has to be controlled along a time trajectory since the operation in batch made is dynamic.

In a continuous plant operation the initial phase or start-up and final phase or shutdown are, however, usually controlled manually since these are the crucial stages where a sequence of steps has to be adhered to strictly. Besides, the dynamic control of a variable is more difficult to implement than steady-state control. The protocol prescribed in these phases has to be followed strictly since their violation can result in an accident.

Large-scale production: The capacity of the modern-day continuous plants is of the order of millions of tons per annum. This is necessary to exploit the *economies of scale*, i.e. it is more economical to run large-scale plants than small-scale continuous plants. This reduces the variable costs of the plant for a unit amount of product formed. An example of a large-scale plant operated continuously is the *Reliance refinery* in Jamnagar. It is one of the largest refineries in the world with a capacity of 1,230,000 bbl/day or 195,000 m^3/day.

In designing a continuous plant the questions asked are as follows.

1. What should be the capacity of the plant? How would you determine this?
   For this, it is necessary to carry out a market survey and find out if there is a demand for the product you intend to manufacture and quantify this demand.
2. How should each unit be designed? Where should the inlets and outlets for each unit be placed? What should be the size of the different units and the flow rates in these units?
   For instance, in a reactor if the inlets and outlets are very close to each other, the reactants may not spend sufficient time in the reactor.

Rather the flow may short-circuit and leave immediately. This would be a poor reactor design.
3. How does one ensure the flow from one unit to the other? Can gravity be used to ensure the flow or do we need pumps?
The former would help us save energy.
4. What kind of provisions must be made to allow heat to be transferred to and from the unit so that the load on the utilities is low?
This could make the operation more energy efficient. It can make the difference and help your organization survive in a fiercely competitive environment.

We now discuss two case studies to illustrate the principles and factors which determine how the technology used in the manufacture of a product evolves with time. These examples serve to highlight the role of the environment, energy efficiency and economics in this evolution.

Now we know that batch processing is used for the manufacture of compounds when the volume of production is low. This is true for speciality chemicals like perfumes, pharmaceutical, products, etc. The continuous production is preferred for large-scale production of industrial chemicals.

Case Study 1: Manufacture of Sulphuric Acid

The industrial development of a country a couple of centuries ago (around the time of the Industrial Revolution) was measured by the amount of sulphuric acid manufactured and consumed. Several factors influenced the evolution and changes in the technology used in the production of sulphuric acid. This example is chosen to demonstrate the factors which influence and drive changes in technology in the manufacture of a product. A good chemical engineer has to understand the underlying principles of a process. This will help him address the challenges faced by the industry. It will enable him to come up with ways to use his knowledge and bring about changes in the manufacturing process to meet the conflicting demands of the society for economic growth and environmental protection.

Lead Chamber Process

The earliest process of sulphuric acid manufacture was the *lead chamber process*. Before the invention of this process, sulphuric acid was manufactured in small quantities in glass bottles. The invention of the lead chamber process made it possible to manufacture it in large quantities literally by the tons. This process was invented by John Roebuck in 1746.

Here potassium nitrate was used as an oxidizing agent to oxidize sulphur to sulphur trioxide. This was then absorbed in water to give sulphuric acid. The reactions which describe this are

$$6KNO_3(s) + 7S \to 3K_2S + 4SO_3 + 6NO$$

$$4SO_3 + 4H_2O \to 4H_2SO_4$$

Let us first see how this process was carried out in the early days. In the original lead chamber process two chemicals, sulphur and potassium nitrate, were mixed and ignited in a room lined with a lead foil. Potassium nitrate or *saltpetre* was used as an oxidizing agent and it oxidized the sulphur to sulphur trioxide according to the reaction given above.

The floor of the room was covered with water. The sulphur trioxide liberated reacted with this water, and sulphuric acid was produced. This process was essentially a batch process and resulted in the consumption of potassium nitrate, an expensive raw material. Besides the NO released also polluted the atmosphere. It also generated solid potassium sulphide which had to be disposed of.

Do you see anything fundamentally wrong with this process, other than the fact that it works? The product *sulphuric acid* contains three elements: hydrogen, sulphur, and oxygen. However, in the process just described, potassium nitrate is consumed to produce sulphuric acid. Specifically, a compound containing nitrogen and potassium is used (consumed) to manufacture something which does not contain either of these elements. Consequently, these elements leave the system as a waste product (potassium sulphide and nitric oxide).

In Great Britain in the early days, potassium nitrate had to be imported from Chile. This was an expensive raw material and it determined the economics of the process. In addition to this, gaseous NO was emitted into the atmosphere and hence the manufacturing process contributed to air pollution. There was, hence, a strong impetus to shift from this process or modify it for both economical and environmental reasons.

Recycle of NO

Joseph Gay-Lussac modified the process by recovering the nitrogen leaving as nitrogen monoxide and recycling it, in 1835. This reduced the consumption and dependence on saltpetre as a source of nitrogen. It made the process more economical and also reduced the impact on the environment. The basic idea can be summarized as follows: nitric oxide was absorbed in water to give nitrous acid. This was reacted with sulphur dioxide to give sulphuric acid, simultaneously releasing the nitric oxide which could be recycled again. The chemical reactions describing these steps can be written as

$$4NO(g) + O_2(g) + 2H_2O(l) \to 4HNO_2(l)$$

$$4HNO_2(l) + 2SO_2(g) \to 2H_2SO_4(aq) + 4NO(g)$$

In the first reaction, nitric oxide is oxidized and absorbed with water to give nitrous acid. This in turn converts sulphurdioxide to sulphuric acid liberating

NO. Thus, we can see that NO consumed in the first reaction is liberated in the second reaction as a product. This accomplished two things simultaneously: it reduced the dependence on expensive saltpetre and at the same time sharply reduced nitrogen monoxide emissions. In the above reactions we can view NO as a catalyst (though its effect on the reaction rate is not being emphasized) as it is not consumed by the overall reaction and only circulates inside the system. It facilitates the production of sulphuric acid without being consumed just like a catalyst. Ideally speaking, we would not have to use any fresh saltpetre or source of nitrate since the NO is continuously regenerated and recycled. Of course, there would be some losses and this would be made up by adding fresh saltpetre or nitric acid. This approach would eliminate the dependency on the saltpetre, thus, making the process economically attractive. The modification suggested by Gay-Lussac, hence, makes the process economical and environmentally friendly.

Figure 2.2 explains the idea of NO recycle proposed by Gay-Lusaac. The SO_2 produced in the sulphur burner is taken to the first column (B). Here it reacts with nitrous acid giving sulphuric acid which is withdrawn from the bottom. The NO released here is absorbed with water to give nitrous acid in the second column (C) which is recycled to the first column.

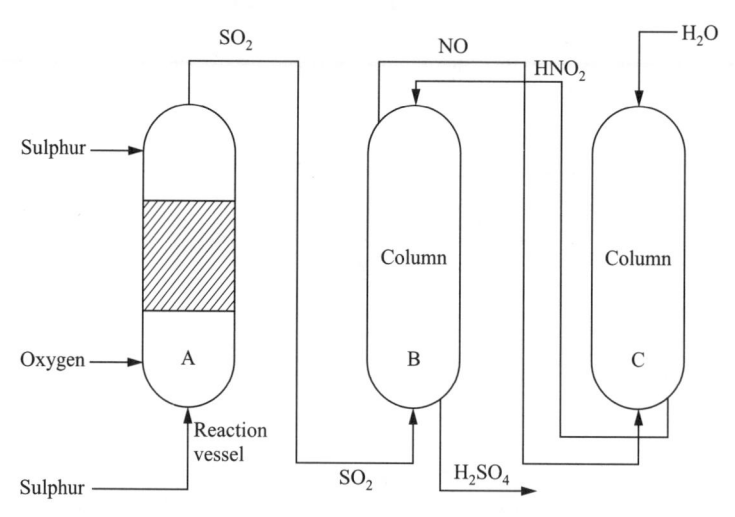

Figure 2.2 Schematic representation of NO recycle proposed by Gay-Lusaac. A-Sulphur burner, B-SO_2 absorption, C-Nitrous acid production.

The overall reactions can be represented as

$$2S + 2O_2 \rightarrow 2SO_2$$

$$2SO_2 + O_2 + 2H_2O \rightarrow 2H_2SO_4$$

While the basic idea behind the process has been explained; let us now see how this was done in an actual chemical plant where sulphuric acid was manufactured in a continuous mode. The classic lead chamber process consists of three units or towers; the Glover tower, the lead chambers, and the Gay-Lussac tower.

A schematic representation of the process is shown in Figure 2.3. It depicts the various units or vessels in which the different processes take place. The streams entering and leaving these units are depicted in the figure. Such a diagram is called a *flowsheet*. Sulphur is oxidized to sulphur dioxide in a furnace or a burner (F). Hot sulphur dioxide gas is sent to the bottom of a vessel called the *Glover tower* (C). Here this gas is scrubbed with nitrous vitriol shown as stream V. This is a concentrated solution of sulphuric acid coming from the Gay-Lusaac tower (D) containing the gases nitric oxide, NO, and nitrogen dioxide, NO_2, dissolved in it. In addition to this weak acid shown as stream W coming from the lead chambers is also used as a scrubbing agent in the Glover tower (C).

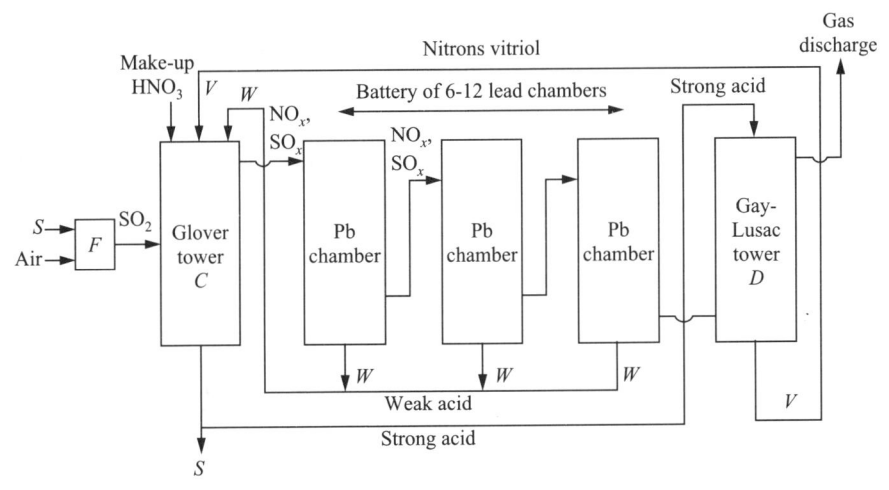

Figure 2.3 Flow sheet of chamber process. V is nitrous vitriol stream, S-Strong acid, W-weak acid.

In the Glover tower (C) two processes occur. First, the weak acid coming out of the lead chamber which is used for scrubbing is concentrated. This is achieved because the SO_2 gas coming from the sulphur burner is hot and the heat helps to evaporate the water from the dilute acid stream. Second, the nitrogen oxides from the vitriol solution coming from the Gay-Lusaac tower are stripped off to the gas phase. As a result of this, the concentration of the acid from the lead chambers increases. Also the NO_x present in the acid coming from the Gay-Lusaac tower is stripped. This latter effect is vital to ensure that we have circulation of NO in the plant.

The acid fed to the Glover tower from the lead chamber's stream W has a concentration of 62% to 68% H_2SO_4. The increase in concentration of this stream is achieved by the hot gases (sulphur dioxide) entering the Glover tower which evaporate water from the acid. Some of the sulphur dioxide is oxidized to sulphur trioxide here and dissolved in the acid wash to form a concentrated solution of about 78% H_2SO_4 called the *tower acid* or *Glover acid* (stream S). The dissolved nitrogen oxides are stripped from the acid and carried with the gas out of the Glover tower into the lead chambers.

From the Glover tower the mixture of gases (including sulphur dioxide and trioxide, nitrogen oxides, nitrogen, oxygen, and steam) leaving is sent to a lead-lined chamber. Here it is reacted with water. Sulphuric acid is formed by a complex series of reactions; it condenses on the walls and collected on the floor of the chamber. There are normally several (3–12) chambers in the plant connected in series. The acid produced in the chambers, often called the *chamber acid* or *fertilizer acid*, contains 62% to 68% H_2SO_4. The reactions occurring in the lead chamber can be represented by an overall reaction.

$$SO_2 + \tfrac{1}{2}O_2 + H_2O \rightarrow H_2SO_4$$

After the gases have passed through the chambers they are sent into another tower called the *Gay-Lussac tower* (D), where they are washed with cooled concentrated acid coming out from the Glover tower (stream S). The nitrogen oxides and unreacted sulphur dioxide are absorbed in the acid to form the nitrous vitriol used in the Glover tower. The waste gases exiting the Gay-Lussac tower are usually discharged into the atmosphere. The stream containing the dissolved NO_x and SO_x (nitrous vitriol) is recycled to the Glover tower.

The product *acid* at a concentration of 78% H_2SO_4 is drawn from the cooled acid stream that is circulated from the Glover tower to the Gay-Lussac tower. Nitrogen losses are made up with nitric acid which is added to the Glover tower.

Some questions which arise are as follows.

1. Why is the dilute sulphuric acid stream W recycled to the Glover tower?
   This is done to concentrate it by evaporating the water. If there is use for the dilute acid the entire amount of dilute acid produced or a fraction of it can be withdrawn as a product stream.
2. Why is the concentrated stream used in the absorption tower, the Gay-Lussac tower?
   The concentrated stream is devoid of NO_x. The NO_x is stripped out in the Glover tower. So the concentrated acid is used as it has a higher absorption capacity. The weak acid is not used since it has NO in it. Since it is coming from the lead chamber, it is likely to be saturated with NO and so it cannot absorb any more NO. In other words, the absorption efficiency would be low, should the weak acid be used.

The flow sheet of the entire process is schematically depicted in Figure 2.3. It shows the different units and the streams coming from one unit to another.

There are several recycle streams in the flow sheet as can be seen. Hence the different units are coupled and integrated to each other. This is a typical feature in almost all chemical process industries. As a result of this coupling, the downstream units can affect the performance of upstream units.

Contact Process

The modifications of the lead chamber process (NO recycle) made the manufacture of sulphuric acid economical and environmentally friendly. Subsequent to this, vanadium pentoxide (V_2O_5) was found to be an effective catalyst which aided the conversion of SO_2 to SO_3. This led to the development of the contact process for sulphuric acid manufacture.

Here sulphur is first oxidized to sulphur dioxide (SO_2) in a burner as in the lead chamber process. This is then converted to SO_3, according to the reaction, we have

$$SO_2 + \tfrac{1}{2}O_2 \leftrightarrow SO_3 + \text{Heat}$$

This reaction is carried out in a reactor which is packed with V_2O_5 catalyst. Catalysts are porous pellets with a diameter of 10 mm. From the stoichiometry applying Le Chatelier's principle, it is seen that the forward reaction is favoured by high pressure since there is a decrease in the number of moles as the reaction proceeds. The pressure used in the reactor is around 20 atm.

The reaction is reversible and exothermic and, hence, the equilibrium conversion is high when the temperatures are low. However, at low temperatures the reaction rate is low. Hence, although we may get a high conversion at equilibrium, the time taken to attain this value is very large. On the other hand, if the temperature is high the reaction equilibrium shifts to the left and formation of products is less. This equilibrium is attained quickly since the rate is high as the temperature is high.

It is, hence, necessary to optimize the temperature of operation in the reactor for such reversible exothermic reactions. One way to achieve this is to start with a high temperature at the beginning of the reactor and slowly decrease it down the length of the reactor so that in the final stage we have a high conversion. The high temperatures in the initial stages at low conversions result in a high reaction rate. The decreasing temperature profile can give an optimal average rate for the reaction. The optimum temperature trajectory can be determined mathematically. The student is used to determining the optimum of a function by setting the first derivative to zero. This gives the value of the independent variable at which the function is maximum or minimum. In the optimal temperature profile problem we are interested in determining an optimal trajectory or a function which varies with axial position. This can be done using "calculus of variations". The optimal profile helps us in combining the advantages of high rate (initial stages) and high equilibrium values (final stages) and we are able to get a good average rate and also a high conversion.

It is difficult to operate a reactor and control it such that the temperature varies along a particular trajectory. This trajectory is approximated by the chemical engineer so that the practical challenge of having to maintain a profile is eliminated. An easy way to approximate the profile is to divide the reactor into several beds or stages (three or four). Each stage is operated adiabatically so that the gases pass through a stage the temperature rises (due to the heat released by the exothermicity). This is easy to accomplish since all you have to do is insulate the reactor. The heat generated in the reaction is removed by passing the gases leaving each stage through a heat exchanger. This results in a drop in the temperature of the gases before they enter the next stage. The heat recovered from the hot gases can be used to generate steam and run a turbine and produce electricity. This process of adiabatic operation with heat recovery is repeated over the different beds. This approach of dividing the reactor into several stages results in a sub-optimal performance (since we have approximated the optimal profile). However, the advantage is it is easy to implement and as engineers we sometimes make a trade-off between going for ideal optimal behaviour to get the best possible outcome and ease of practical implementation.

The gases leaving the last bed contain unreacted sulphur dioxide and trioxide. The SO_3 is absorbed in a solution of sulphuric acid. This is necessary since when SO_3 is absorbed in water the heat liberated is very high. Further, there is a significant loss in the form of mist which is very corrosive. When it is absorbed in sulphuric acid solution the amount of SO_3 absorbed is lower and so the heat liberated during the absorption (heat of absorption) is lower. Here oleum is formed ($H_2SO_4(SO_3)$). A schematic diagram of this flow sheet is shown in Figure 2.4.

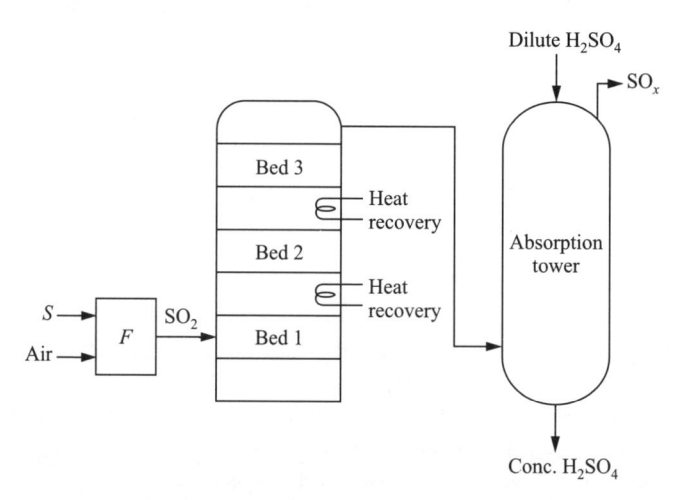

Figure 2.4 Contact process for sulphuric acid manufacture. The reactor has three adiabatic beds with intermediate cooling.

The process which uses the solid V_2O_5 catalyst is the *contact process*. A drawback of this process is that the emissions of unreacted SO_2 from the absorption tower are high. To control this, gases leaving the absorption tower consisting primarily of SO_2 are sent back to the last stage of the multistage convertor. Here the temperature is low and so maximum conversion can be achieved. Here they get converted to SO_3 and the gases leaving are absorbed again in a second absorber. This is the double-contact, double-absorption (DCDA) process used in the industry. A flow sheet of DCDA process is shown in Figure 2.5.

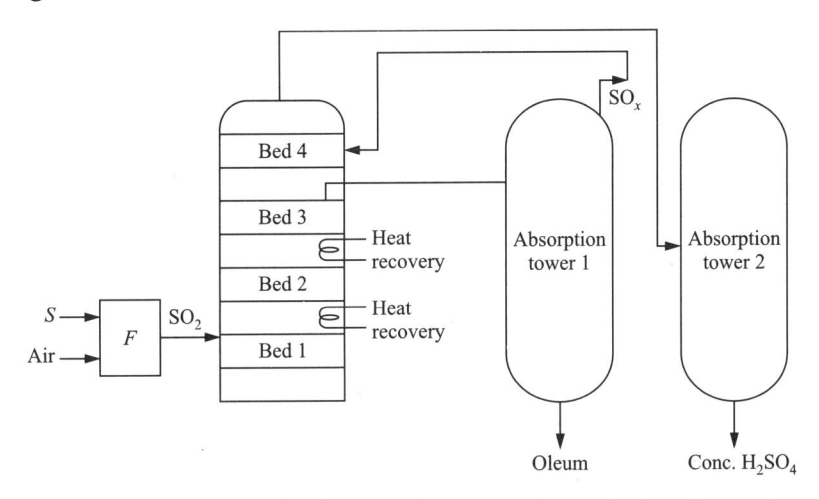

Figure 2.5 Double-contact, double-absorption process for sulphuric acid manufacture.

The multistage reactor with intermediate cooling is a practical way of approximating the optimal profile. Science helps us in determining the optimal profile, but practical issues drive the engineer to innovate and approach the optimal solution taking into account considerations like ease of operation, economics, and plant safety.

Implications of Coupling and Recycling: Start-up and Shutdown

In view of the integration and coupling across the different units in a typical flow sheet as the lead chamber process it is difficult to see how the process is started. The flow sheet represents the steady-state situation of the plant. The steady-state is the state attained after the plant has been in operation for a long time. Here all the flow rates and the variables like concentrations and temperatures at any point in the plant are constant and do not change with time. How does one reach the steady state? For instance, when you start the plant you would generate SO_2 in the furnace but there would be no recycle streams coming to the Glover tower from the lead chamber or the Gay-Lusaac tower. Since the flow sheet has recycle streams it is difficult for us to identify the sequence of steps to be followed to reach the steady-state operation in this plant.

To understand this better we must know that in the operation of a process plant there are three stages: start-up, steady state, and shutdown. The process flow sheet represents the flow of streams from one unit to another and the conditions in various units at steady-state conditions. However, the first phase, start-up and the last phase, shutdown are the dynamic states where we have variables (the flow rates in the streams, their compositions and temperatures) changing with time. These are the most crucial stages in a process plant operation. For the most part they are manually controlled. Most of the accidents, problems, etc. occur primarily in these stages. In these phases there is a protocol to be followed, for instance, for the sulphuric acid plants it could be: Turn on the furnace and allow it to reach a fixed temperature. Then another unit is started. The sequence of steps for start-up depends on the particular plant being considered. The duration of the start-up stage can vary from a few hours to a few days. The start-up may involve keeping the different flow rates low at the beginning and then slowly increasing the flow rates to reach the full capacity. Or the products formed in the start-up period may not be withdrawn but be recycled to minimize the waste. Alternatively, the products formed in the start-up may be of a poor quality and may be discarded. A wrong start-up strategy will result in poor performance of the plant. It may end up producing an undesirable product or damaging the equipment. Most problems occur here due to operator error arising from carelessness or the operator being in a hurry and not paying proper attention to the protocol.

Once the steady state is reached, then the plant is usually run on auto control and manual intervention is minimal.

Similarly, shutdown is a procedure where certain steps have to be adhered to sequentially. This may take around the same duration as the start-up procedure. In a chemical plant these processes are very intricate. When you work on a computer and you want to start it or shut it down, you press the power switch. The point is that the chemical plant is more complex and it cannot be started up or shut down by operating just one switch.

In the flow sheets shown in Figures 2.4–2.6, we have not given any quantitative information about the flow rates and the concentrations, temperatures of the streams and various units. To determine this we need to make mass balance calculations after understanding the processes in each unit. The actual fraction of the product stream being withdrawn from the system, and the flow rates of various streams are obtained from mass and energy balance calculations around each unit of the process flow sheet.

Case Study 2: Soda Ash (Sodium Carbonate) Industry

In the early 1800s several halophytes, i.e. salt tolerant plant species and seaweed species were used to produce an impure form of sodium carbonate. The land plants (typically glassworts or saltworts) or the seaweed (typically Fucus species) were harvested, dried, and burnt. The ashes that were left behind were then washed with water to form an alkaline solution. This solution was boiled dry to create the final product, which was termed *soda ash*. This name comes from the primary plant source for soda ash, the small annual shrub *salsola soda*.

The sodium carbonate concentration in soda ash varied very widely, from 2–3% when derived from seaweeds, to 30% when obtained from saltwort plants in Spain. These sources were found to be inadequate with rapid industrialization. This resulted in an intensification in the search for commercially viable routes to synthesize soda ash from salt and other chemicals.

Leblanc Process

In 1791, French chemist Nicolas Leblanc patented a process for producing sodium carbonate from salt, sulphuric acid, limestone, and coal. First, sodium chloride was treated with sulphuric acid to yield sodium sulphate and hydrogen chloride in the gaseous state.

$$2NaCl + H_2SO_4 \rightarrow Na_2SO_4 + 2HCl$$

The sodium sulphate, thus, produced was blended with crushed limestone (calcium carbonate) and coal (essentially carbon). This mixture was burnt, producing calcium sulphide and sodium carbonate.

$$Na_2SO_4 + CaCO_3 + 2C \rightarrow Na_2CO_3 + 2CO_2 + CaS$$

The sodium carbonate was extracted with water, and then concentrated by allowing the water to evaporate.

The hydrochloric acid produced by the Leblanc process was a major source of air pollution, and the solid calcium sulphide by-product also presented waste disposal issues. However, it remained the major production method for sodium carbonate until the late 1880s when the Solvay process was born.

Solvay Process

In 1861, Belgian industrialist and chemist Ernest Solvay developed a method to convert sodium chloride to sodium carbonate using ammonia. The *Solvay process* centred around a large hollow carbonation tower (Figure 2.6). At the bottom of the tower, calcium carbonate (limestone) was heated. This decomposed and carbon dioxide was released according to

$$CaCO_3 \rightarrow CaO + CO_2$$

44 *Introduction to Chemical Engineering*

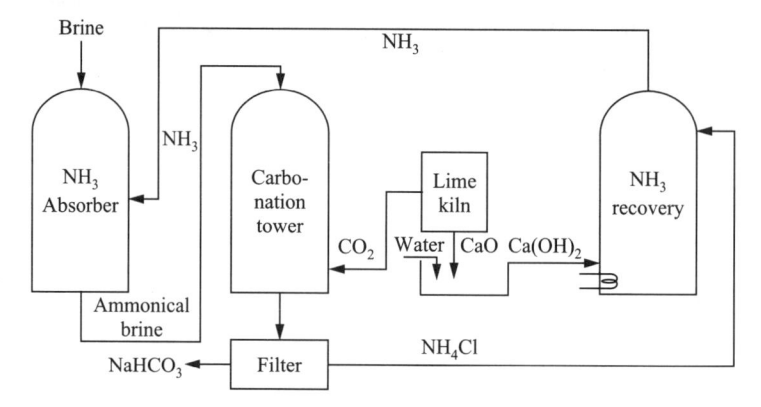

Figure 2.6 Flow sheet of the Solvay Process.

At the top of the tower, a concentrated solution of sodium chloride and ammonia is fed. As the carbon dioxide bubbled up through this liquid flowing in a counter current mode, sodium bicarbonate was formed according to the following reaction

$$NaCl + NH_3 + CO_2 + H_2O \rightarrow NaHCO_3 + NH_4Cl$$

The sodium bicarbonate precipitates out. This was then converted to sodium carbonate by heating it, releasing water and carbon dioxide

$$2NaHCO_3 \rightarrow Na_2CO_3 + H_2O + CO_2$$

The ammonia was regenerated from the by-product ammonium chloride by treating it with the lime (calcium hydroxide) left over from carbon dioxide generation. This can be described by the following reactions.

$$CaO + H_2O \rightarrow Ca(OH)_2$$

$$Ca(OH)_2 + 2NH_4Cl \rightarrow CaCl_2 + 2NH_3 + 2H_2O$$

Because the Solvay process recycles its ammonia, it consumes only brine and limestone, and has calcium chloride as its only waste product. This made it substantially more economical than the Leblanc process, and it soon came to dominate world sodium carbonate production. By 1900, about 90% of sodium carbonate was produced by the Solvay process, forcing the last plant operating on the Leblanc process to close in the early 1920s.

While the chemistry of the process has been described, the chemical engineer designs the reactor, the positions of inlets and outlets of the different streams. For a smooth operation these must be designed so that the precipitates of the bicarbonate do not hinder the flow of the streams or clog the pipes.

The examples of sulphuric acid and soda ash manufacture have been chosen to illustrate how the technology used in their manufacture has changed with the time. The existing technology used today to manufacture a chemical may be rendered obsolete in the future. This can be brought about by environmental or economic considerations.

Common Features between the Evolution of the Sulphuric Acid Industry and the Soda Ash Industry

The technology used in the early days in both processes was environmentally polluting. In the sulphuric acid case (lead chamber process), NO_x was emitted and in the soda ash industry (Leblanc process) HCl gas was liberated. These gases caused pollution. The two processes were both expensive and used materials containing elements which did not occur in the final compound. Potassium nitrate containing nitrogen was consumed in the lead chamber process. Nitrogen did not occur in sulphuric acid. The Leblanc process used sulphuric acid when the final product did not contain any sulphur. Both processes—the lead chamber process and the Leblanc process—had a problem with solid waste disposal, potassium sulphide, and calcium sulphide.

The two processes were modified to ensure that the raw material which was unnecessarily consumed (potassium nitrate and sulphuric acid) could be eliminated. Thus, in the sulphuric acid process NO was used as the catalyst or carrier while in the soda ash process ammonia was used as the catalyst or carrier gas. This made the two processes economical and environmentally friendly.

A good chemical engineer is one who understands the scientific basis for the changes brought about in technology of production. He should also realize the similarities in apparently dissimilar processes. This helps illustrate that the same basic principles have applications in several diverse areas and applications. We have seen the similarities in the evolution of the sulphuric acid and soda ash industry. The application of dialysis and effluent treatment is also intrinsically similar as explained in Chapter 1.

Next, as another example of manufacture of different chemicals which operate on a common principle, processes with recycle streams is considered.

Processes with Recycle Streams

Many processes are based on reactions which are intrinsically slow or are limited by thermodynamics. Here the conversion of the reactants is not more than 15–20% in the reactor. The raw materials used in these processes could be expensive and hence in order to render the process economical it is necessary to separate the products from the reactants in the stream leaving the reactor. For example, in the manufacture of ammonia the conversion realized is around 15%. Nitrogen usually comes from an air liquefaction unit and hydrogen from a steam reforming unit. Hence, both reactants are expensive since their production is energy intensive. It is, hence, necessary to recycle the un-reacted reactants to the reactor in order to make the process economical.

After the separation in a downstream separator the unconverted reactants are recycled back to the reactor. If we do not recycle the unconverted reactants, then we would be venting out expensive reactants and the efficiency of the process would be low and the cost of the production would increase.

The principle of reactant recycle is the basis of many chemical process industries. Systems characterized by low conversions in the reactor occur in the manufacture of ammonia, methanol, acetic acid, and low density poly ethylene (LDPE). In these cases the reactions which govern the product formation are reversible reactions. Hence, the conversion obtained is governed by equilibrium conditions and this can often be low.

In the manufacturing of these compounds fresh feed containing reactants is added to the reactor. Separation of the products from the reactants is achieved in a downstream separator. The product-rich stream is taken downstream for further processing while the reactant-rich stream is recycled back to the reactor. This recycle ends up ensuring a two-way coupling between the reactor and the downstream separator unit and hence the behaviours of the two systems are interlinked. The reactor and the separator unit depend on the particular process chosen. In ammonia manufacture the separator is a refrigeration unit while in acetic acid manufacture it is a distillation column. A schematic diagram depicting this recycle stream coupled reactor-separator is shown in Figure 2.7.

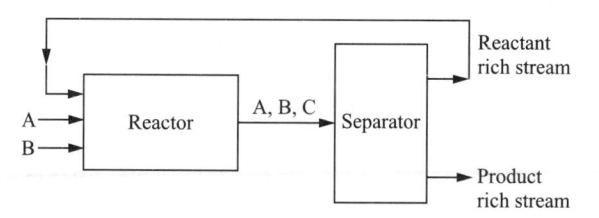

Figure 2.7 Coupled reactor-separator systems.

Note to the Instructor on Plant visits: One of the primary responsibilities of a chemical engineer is to be able to depict a process in the form of a flow sheet. The flowsheet of a chemical process plant is usually very intricate and shows the close interlinkages the different units have in a process. The ideal means of exposure to this is to take the students to a process plant. The students have to understand the process and draw the flow sheet. The choice of the plant is important. It should not be too complex; otherwise the student may get lost in details and feel intimidated.

Reverse Osmosis Plants

Most colleges have reverse osmosis or similar water treatment facilities on their campus which provide drinking water to the students. The IIT Madras has two plants based on the principle of reverse osmosis (RO) to purify water. This purified water is supplied to the students in their hostels (dormitories) for drinking. The feed to the plant is either well water or the water supplied by municipal authority. The students are divided into small groups and each group visits the plant. The different units are described. The processes occurring in them and the principles on which they act are explained. A schematic diagram of the plant is shown in Figure 2.8.

Modern Chemical Engineering Plants 47

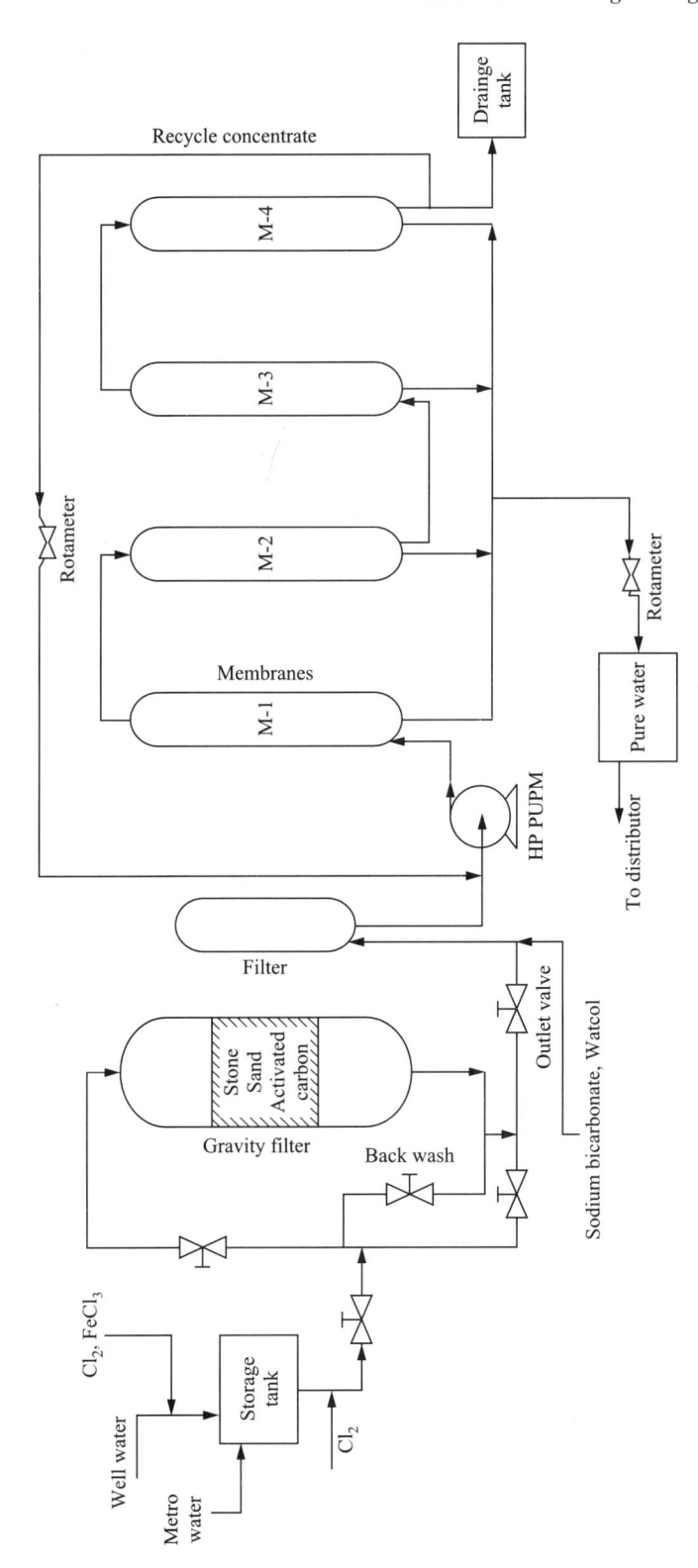

Figure 2.8 Flow sheet of RO plant in IIT Madras.

The plant visit would help the student understand the level of detail to which the engineer has to cater to. In the classroom the concept of reverse osmosis can be explained, where it is expounded that the pressure has to be more than the osmotic pressure for the solvent to move from the high solute concentration region to the lower solute concentration region. The addition of coagulants, the pre-treatment of water using chlorine and other chemicals and their importance can be best understood only when they see the process in operation. The effect of high chlorine and the possibility of membrane fouling and the precautions which need to be taken to address this are best understood by a plant visit. The need to have a backup plan in case of failure of units and redundancy of units when critical operations are involved can be explained in the class but that is not as effective as a plant visit.

A common method used for desalination of seawater is reverse osmosis. It can also be used to treat brackish water so that the concentration levels of total dissolved solids are lowered to acceptable levels making the water potable. The energy consumption in an RO process is significantly lower than in a multiple effect evaporator and hence RO is usually preferred.

Water purification is a process of removing undesirable chemicals (soluble and insoluble) and biological contaminants so that they do not have any adverse impact on our health. The goal is to produce water fit for a specific purpose, i.e. make water potable, or for medical, chemical, and industrial applications. The level of purification required is decided by the applications of water.

There are several steps used in the water treatment. The first step is the primary treatment where physical processes such as *filtration* and *sedimentation* are used to remove the suspended particles. In the secondary treatment, an activated sludge process is used where microorganisms treat the wastewater biologically. In the tertiary treatment step, chlorine or ultraviolet light is used to treat water.

The purification process is aimed at reducing the concentration of particulate matter including the suspended particles, parasites, bacteria, algae, viruses, fungi, and a range of dissolved materials present in the contaminated water.

Typical Processes Involved in the Plant

Rapid Gravity Filter

Filtration is a physical operation which is used for separation of solids from fluids (liquids or gases). The mixture flows through a porous bed through which only the fluid can pass. Oversized solids in the fluid are retained in the bed. Here the separation is usually not complete as the filtrate will contain fine particles (depending on the pore size and filter thickness). In the plant, rapid sand filters containing relatively coarse sand and other granular media are used to remove the particulate matter. In addition to the particulate matter, it can also remove flocs. Flocs are formed using flocculating agents which are typically salts of aluminimum or iron. They have impurities trapped in them. Water containing flocs flows through the filter medium under gravity or under pressure and the flocculated material is trapped in the sand matrix. Ferric chloride is typically used as flocculating agent.

Disinfection

Disinfection is the process in which microorganisms like bacteria and viruses are killed using oxidizing agents. These agents act by oxidizing the cell membrane of microorganisms. A large number of disinfectants cause cell lysis and death. Chlorine is typically used in most systems.

Reverse Osmosis

Reverse osmosis (RO) is a filtration method that removes many types of large molecules and ions from solutions by applying pressure to the solution when it is on one side of a selective membrane. The result is that the solute is retained on the pressurized side of the membrane and the pure solvent is allowed to pass to the other side. This is the reverse of the normal osmosis process, in which the solvent naturally moves from the area of low solute concentration, through a membrane, to the area of high solute concentration. The movement of the solvent to equalize solute concentrations or chemical potentials on each side of a membrane is visualized as arising out of pressure which is generated and this is called the *osmotic pressure*. Applying an external pressure to reverse the natural flow of pure solvent, thus, is reverse osmosis. Typically, polymeric membranes are used. These polymeric membranes are made up of cellulose acetate, poly ethylene, poly propylene, polysulfones, PTFE, etc. Several shapes such as spiral wound membranes are used in commercial plants.

We will now describe how these processes are implemented in the RO plant in IIT Madras (Fig. 2.8). This gives us an idea of the planning an engineer has to do to bring a concept to fruition using the basic principles of each unit.

Description of the Process of RO Plant in IIT Madras

1. The feed water to the RO plant is from the municipal supply as well as the underground water from the wells. This feed water is taken to a storage tank consisting of two compartments, each of a capacity of 50,000 litres. It has two inlets, one for well water and the other for municipal water. The water is allowed to settle for around 6 hours after adding ferric chloride ($FeCl_3$) as a flocculating agent in the

storage tank. This induces agglomeration where smaller size particles coalesce with each other and form large particles and settle down at the bottom. The clear water from the storage tank is sent to the gravity filter section. The water from each compartment of the storage tank is used alternatively as the feed to this unit. The water is fed to the top of the gravity filter through an inlet valve.
2. The gravity filter consists of three consecutive beds. Each bed is packed with a different material. The first bed is made of stones, the second of sand and the third of carbon. The water flows through each of these beds under the action of gravity.
3. The stone layer is a coarse bed with high porosity. It removes larger diameter suspended particles and allows the water to pass through. In the layer of sand, particles of medium size are removed, and finally in the final bed of carbon smaller size particles are removed.
4. The treated water coming out of the gravity filter is sent to the chlorine detector section. If the chlorine levels are found to be high, then sodium bislphite is added to neutralize the excess chlorine, chlorine levels have to be controlled in the water as this can degrade the membranes in the RO unit. The addition of chlorine is necessary since it is a disinfectant. However, it should not be in excess. The water is also checked for other impurities. If impurities are found to be high, then it is recycled through the recycle valve back to the gravity filter. This is the extra precaution taken to ensure the membrane of the RO unit does not get clogged. This is achieved by opening and closing some valves.
5. The treated water is sent to the next filtration unit where a further purification is carried out. This unit consists of a hollow cylindrical pipe surrounded with a fibrous material. The treated water from the gravity filter enters the inner surface of the filter, flows radially outward across the filter. This helps remove any suspended particles which have not been removed in the gravity filter.
6. The water coming out of this unit is pumped to the first column (M1) of the reverse osmosis unit from the bottom, where the soluble impurities get removed. The purified water from the first column now passes through the second column (M2) and then to the third (M3) and, finally, to the fourth column (M4). The purified water from the reverse osmosis unit is sent to the storage tank. The flow rate of the pure water is measured using a *rotameter*. The high solute concentration water from the reverse osmosis column (the retentate) is sent either for recycling or to the drainage tank.

In this process, the following chemicals are used:
(a) Chlorine is used as a disinfectant to kill microorganisms.
(b) Ferric chloride is used as flocculating agent.
(c) Sodium bisulphate is used to remove excess chlorine from water.

While the principles of the process can be explained in the class, the visit to the plant helps understand some finer points. These include:
1. With the passage of time the gravity filter bed gets clogged with impurities and flocs. This will have to be removed. A typical way to acheive this is to pump the water in the reverse direction. This dislodges the trapped particles and reactivates the bed.
2. There is a provision to verify the quality of the water after the gravity filter. Should the quality be found to be poor, the water is sent back to the gravity filter through a recycle line. These steps are extremely important as they prevent the fouling of the membrane in the RO unit.
3. The membrane used in the RO unit may be sensitive to chemicals present. It is, hence, necessary to check the level of these chemicals like chlorine before the water is sent to the RO unit.

This kind of detailed planning is necessary to ensure that the plant can operate continuously without any interruption for a long period of time. A flow sheet describing the RO process in IIT Madras is shown in Figure 2.8.

Exercises

1. What are the basic questions you would ask when you have to design a plant? How would you decide the capacity of the plant?

   Once the capacity of a plant is determined, material balances and energy balances determine the concentrations, temperature and mass flow rates of various streams in the process. Then we have to design the various equipment, i.e. determine the size, etc. of each unit, the material of construction, etc. This is done in specific courses of curriculum later.

2. List some specific chemicals manufactured by batch processing where chemical reactions are involved.

3. Describe the process of manufacture of one of the chemicals in #2 in detail.

4. How is NaOH manufactured today? How was it manufactured earlier? What caused the shift in the manufacturing process?

5. Explain the similarities in challenges and changes in the process of manufacture of sulphuric acid and soda ash.

6. What is the kind of reactor used in the manufacture of sulphur trioxide from sulphur dioxide? Explain why such a reactor is required.

52 Introduction to Chemical Engineering

Appendix—Liquid Nitrogen Plant

The Plant

The liquid nitrogen plant is housed in the Physics Department of IIT Madras. It has the capacity to produce 25 L/h. Liquid nitrogen is used for carrying out research in the area of cryogenics or low temperature physics. It is also used as a coolant for several sophisticated instruments like the high resolution scanning electron microscope, nuclear magnetic resonance imaging, high and ultra high vacuum systems, preserving biological samples, and for activating chemical reactions.

The Process

The process in the IIT Madras nitrogen liquefier begins by filtering the air from the atmosphere to remove the particulate matter. The filtered air is sent to a screw compressor. Its main components are the housing and two rotors. The compressor operates based on the rotary piston principle and is driven by an electromotor through a V-belt. During the compression the pressure is raised to 8 bar. Consequently, the temperature of the air also rises (as can be seen from the equation of state).

The hot compressed air is fed into a refrigerated air dryer unit which removes the moisture from the compressed air. This is accomplished by cooling the air with a refrigeration unit to below the dew point causing the moisture in the air to condense. The minimum air temperature that can be obtained using this unit is 2°C. This cooling is done to condense any water vapour present in the compressed air. The condensed water is removed in the form of liquid water using a solenoid valve which is opened periodically. To prevent the formation of ice in this unit ethylene glycol is added to the water.

After this, the air is fed to an oil filter and the pressure swing adsorption unit. Here the oxygen is separated from nitrogen. This is done by using carbon molecular sieves which selectively adsorb oxygen molecules which are smaller (4.36 Å) as compared to nitrogen (4.64 Å) and allow nitrogen to pass through. The pressure swing adsorption consists of two cylindrical columns filled with molecular sieves. Regular regeneration is achieved with two adsorption vessels filled with activated carbon sieves that operate alternatively. The adsorption cycle of each column has a duration of around 60 seconds. At the same time, while in one vessel, oxygen is removed from the incoming air, the internal pressure of the other vessel is released to the atmosphere to get rid of the gases trapped during the previous cycle. Since the pressure swings alternatively from one sieve vessel to the other, the technique is known as *pressure swing adsorption* (PSA). The produced nitrogen (99.98% pure) leaving the adsorption vessel flows into the buffer tank. This buffer tank ensures constant supply of nitrogen to the cryogenerator. The cryogenerator uses a 'Stirling' refrigeration cycle and provides cooling to the re-condenser. It generates a low temperature (60 K) in the condenser unit with helium as a working gas. The cryogenerator requires a continuous supply of cooling water to support the Stirling cycle and to cool the oil contained in its lubrication system. Gaseous nitrogen is pumped into the re-condenser at a pressure of up to 2 bar. The non-condensable gases are exhausted to the ambient through a mini vacuum pump. The liquid nitrogen produced in this process is drained into a storage tank (2000 litres capacity) through the re-condenser outlet. This tank is maintained at a pressure of two bars. Some nitrogen gets converted to vapours. These vapours are recycled back into the cryogenerator. Figure A1.1 shows the flow sheet for liquefaction of N_2. Figure A1.2 presents a photograph of the liquid nitrogen plant in IIT Madras.

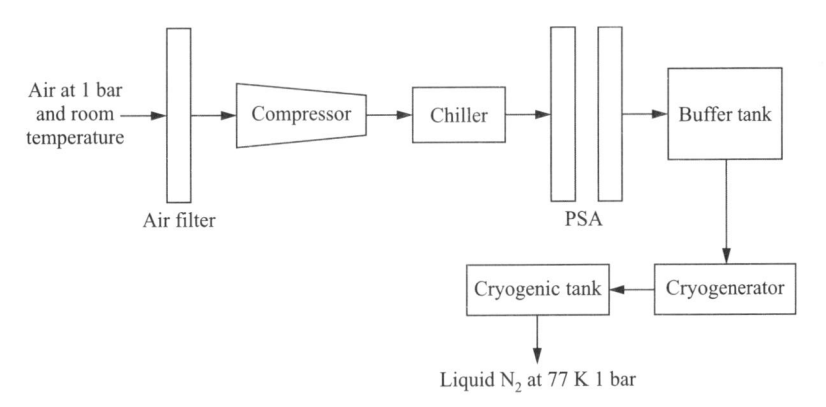

Figure A1.1 Flow sheet for liquefaction of N_2 in IIT Madras.

Figure A1.2 Liquid nitrogen plant.

3

Chemical Engineer and Chemical Engineering Profession

Introduction

The chemical engineering profession is a relatively young profession, less than 100 years old. Manufacturing of chemicals at various scales or plant capacities, however, has been going on for a much longer period. In the early days before the formal chemical engineer was born, mechanical engineers with a good knowledge of industrial chemistry were responsible for the operation of chemical plants. An engineer working in a urea plant, for example, had enough knowledge of the chemistry of reactions involved in the manufacture of urea and the operations of the different pieces of equipment of the urea plant. He understood the processes occurring in various units of his plant. He could operate and maintain it. However, he did not have the expertise or capacity to work in any other plant say, for instance, a sulphuric acid plant. By the same token a person working in a sulphuric acid plant, for instance, would be completely lost in an ammonia plant since as far as he was concerned the reactions and processes in different units of the two plants were different. The knowledge of the mechanical engineer was restricted to the processes, the systems and operations of the plant he was working in. He did not have the foundation or basic knowledge or training which could give him confidence to work in another chemical plant. The focus in the early days was on individual technologies and not on unification of principles. Thus, students would get trained in the manufacture of a particular chemical but would not have basic concepts to solve problems arising in another industry.

The above situation is similar to the following. Consider a student A who memorizes various methods to solve different mathematical problems without really understanding the principles behind the methods used. Consequently, if a student using this kind of an approach were to be given a different but "similar" problem he may not be able to find the solution since he has not understood the fundamental principles which lie behind seeking a solution. He is unable to solve it since he is not used to solving that specific problem. However, if another student B has a fundamental understanding of the methods of solution he uses, then he is versatile enough to solve any problem given to him which is based on the same principle although the problem may look seemingly different. The student A is similar to the mechanical engineer running a chemical plant. This student is not versatile enough to work in any other plant. The student B is similar to the chemical engineer who can run any chemical plant.

The similarities that exist between the apparently dissimilar processes of haemodialysis and ultrafiltration have already been pointed out in Chapter 1. The similarity in the two processes is easily seen and understood by a chemical engineer because of his training. There was hence the need for the profession to train people to work in chemical process plants by understanding the basic similarities in the processes taking place in the various units. A chemical engineer has a good fundamental understanding of the various processes occurring in a plant. This gives him the necessary skills to work in different chemical process industries which operate under the same basic fundamental principles.

The Birth of Chemical Engineering

George Davis (1850–1906) is considered the *father of chemical engineering*. Davis studied at the Slough Mechanics Institute and the Royal School of Mines in London (now a part of Imperial College, London). He worked in chemical industries around Manchester. Before he embarked on a career as a consultant, he held various positions—one as an inspector for the Alkali Act of 1863. This was a very early piece of environmental legislation that required soda manufacturers to reduce the amount of hydrochloric acid gas vented into the atmosphere from their factories. His job profile was such that he had to visit various chemical plants and inspect their operations. During the course of these visits he found several similarities in the processes occurring in various units of different plants. He made a comprehensive study of the different processes in these plants and highlighted the fundamental principles on which these processes were based.

In 1887, Davis gave a series of 12 lectures at the Manchester School of Technology, which formed the basis of *Handbook of Chemical Engineering*. At that time there were already several industrial chemistry books written for each chemical industry—for example, alkali manufacturing, acid production, brewing, and dyeing. Davis's contribution was that he organized his text by the basic operations common to many industries—transporting solids, liquids, and gases; distillation; crystallization; and evaporation, to name a few. His lectures were criticized as being common place *know-how* since these were designed around operating practices used by British chemical industries.

The observation of the fundamental similarity in the different processes in the various plants led to the introduction of the concept of unit operations. This concept provides a scientific basis and a unified approach for understanding the behaviour of processes in the various units; reactors, separators, etc. in apparently different chemical plants. This approach helps us in providing a unified framework for understanding and describing processes operating on similar principles. Adopting this approach in the curriculum helped develop confidence and versatility in the students to work in different chemical plants. This shifted the emphasis from a technology-based approach to a science-based approach. Let us see an example of such an underlying similarity in two dissimilar processes.

Distillation

As an example of a unit operation we discuss "distillation". Consider (i) the separation of a mixture of ethylbenzene and styrene into pure components, and (ii) the concentration of sulphuric acid from a dilute solution of the acid. These two processes at the outset look completely different. However, they are a separation technique exploiting the difference in boiling points of the individual components (ethylbenzene and styrene in the first case and water and sulphuric acid in the second case), i.e. they are based on the same principle. When a mixture containing two components boils the vapour phase is richer in the more volatile or "lighter" component whereas the liquid phase is richer in the less volatile or "heavier" component. Specifically if we have a mixture of methanol (boiling point 333 K) in water (boiling point 373 K) and heat the mixture to a temperature of 353 K the vapour phase will be primarily rich in methanol whereas the liquid phase will be rich in water. You may be tempted to think that all the methanol would come to the vapour phase leaving only water in the liquid phase. You will see later on in your course on "Mass Transfer" that both phases will have both components. The relative amount of the various components in each phase will, however, be different. Thus, in the separation of ethylbenzene and styrene as well as in the concentration of sulphuric acid the difference in boiling points is exploited. The unit operation where separations are based on exploiting the differences in boiling points is called *distillation*. This forms a separation technique and is discussed extensively in the course on Mass Transfer. A schematic diagram of this is shown in Figure 3.1. Here V represents the vapour stream and L the liquid stream. The two streams flow counter-current and the lighter (heavier) component is withdrawn from the top (bottom). Here F, D, W represents the flow-rates of the feed, distillate and residue streams. R is the reflux or fraction of D being recycled.

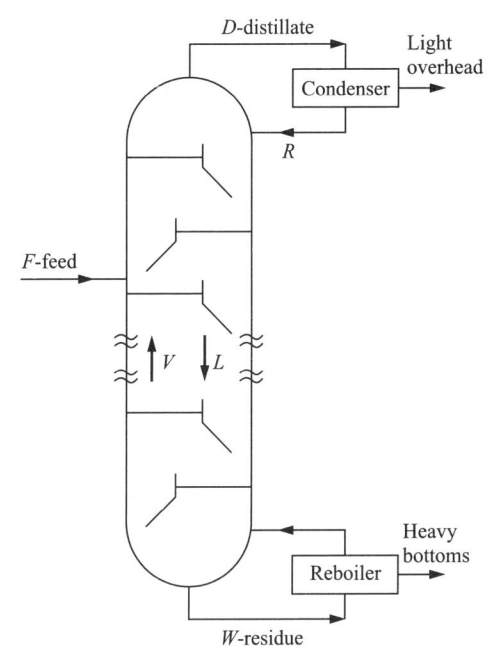

Figure 3.1 Schematic representation of a distillation process.

We now come back to the history of chemical engineering. The concept of unit operations was based on the unification of principles governing the behaviour of processes in different systems. Arthur D. Little pioneered that the core of chemical engineering education should be centred on the "Unit Operations", a study of the steps common to most industrial processes such as fluid flow, heat transfer, distillation, filtration, crushing, grinding, and crystallization. Paradoxically, all the unit operations refer to physical rather than chemical operations in the chemical industry.

At the core of any chemical industry lies the chemical reactor where the reaction occurs and reactants are transformed to products. These reactions were grouped into unit processes (Groggins, 1935). However, this was mainly limited to organic reactions, alkylation, esterification, nitration, sulphonation, etc. This was soon extended to inorganic processes (Shreve, 1950), providing a rather comprehensive and unified framework of all processes in the chemical industry. This helped perform a comparative study of chemically reacting processes. In spite of this unification the emphasis in this phase was, however, more on empirical knowledge and applications of ideas obtained from practice. This went on well till the 1960s.

Susbsequent to this there was a drive to obtain a scientific understanding of basic principles governing the processes in different units. The motivation for this was to reduce the dependency on empiricism. The classical book *Transport Phenomena* by Bird, Stewart and Lightfoot was the first step in this direction. It presents systematically and interrelates momentum transport (fluid mechanics), energy transport (heat transfer), and mass transport (diffusional processes). The discussion here is at three levels: molecular (transport properties), continuum (equations of change), and equipment (macroscopic balances).

The study of transport phenomena provided the unit operations with an advanced scientific basis in the language of physics and mathematics. While dealing with the unit operations it endowed the engineer with greater ability to innovate and to create from the first principles. Moreover, the integration of transport phenomena with chemical reactions led to chemical reaction engineering (Levenspiel, 1962). This is devoted to the selection, design, and operation of chemical reactors.

Armed with the knowledge base of unit operations, transport phenomena and chemical reaction engineering, the chemical engineer became highly versatile and proficient in fields other than the namesake of the profession, the chemical industry. He could be involved in petroleum, metallurgy, materials, biology, agriculture, pharmaceuticals, medicine, environment, combustion, information technology, etc.; some of these are not chemical fields.

To better describe such a broad spectrum of activities, the classical name of chemical engineering has been extended to process engineering, as a generic designation to describe all processes involving the treatment of materials, be that physical, chemical, or even biological (enzymatic).

At this point, the unit operations (UO), the principles of transport phenomena (TP), and chemical reaction engineering (CRE), taken together, had become the unique knowledge base of chemical engineering. This resulted in the large-scale production of commodity chemicals through the design, optimization, and operation of chemical plants.

This dissolution of boundaries between disciplines opened the gates to chemical engineers to orient themselves more towards molecular sciences and mathematical techniques. It gave confidence to chemical engineers to design molecules leading to the production of "speciality chemicals" and produce them on a plant scale. The performance of these high value, speciality products, *e.g. antibiotics and other pharmaceuticals and functional materials* depend on their chemical composition and purity and also on the geometrical structure on the molecular scale.

Curriculum

We now discuss briefly the various courses in the chemical engineering curriculum and the concepts they deal with. An overall picture is presented for each course by way of motivation and these are not discussed in detail. The chemical engineering curriculum consists of courses on *Momentum Transfer, Heat Transfer, Mass Transfer, Thermodynamics,* and *Chemical Reaction Engineering*. The courses on Mass Transfer and Chemical Reaction Engineering are the ones which are unique to the chemical engineer and gives him the versatility to work across industries. This also makes the chemical engineer more versatile and interdisciplinary enabling him to work across diverse disciplines and collaborate with mechanical engineers, biotechnologists, etc.

Let us now look at how these courses help a chemical engineer and the kind of questions that are answered in these courses.

Thermodynamics

Rate processes (heat, mass, and momentum transfer) and thermodynamics are two different but fundamental aspects of engineering. Consider a steel rod such that the half of the rod is at 80°C and the other half is at 140°C. If the rod is insulated, thermodynamics would tell us that the final temperature attained by the rod would be uniform at the average value of 110°C. The transfer of heat from one end to the other would be by conduction till the temperature gradients vanish. However, thermodynamics does not give us information on how quickly this uniform temperature is going to be attained. For this we need to understand the rate of heat transfer from one half of the rod to the other half.

In designing the systems discussed above where we talked about distillation and extraction thermodynamics plays an important part in computing the equilibrium compositions of the two phases arising in a separation process. For distillation these phases are the vapour and liquid phases while for extraction these phases are the two liquid phases called the *extract* and the *raffinate*. Consider an aqueous solution of acetic acid in a beaker. On addition of kerosene to this solution and mixing, it will result in a portion of acid moving from the aqueous phase to the organic phase. After sufficient time has elapsed the concentration in both phases will be a constant. These are the equilibrium compositions. The equilibrium composition is the composition of the two phases when sufficient time has lapsed and no more transfer of species can occur from one phase to another. The composition does not change beyond the equilibrium value when the two phases are in equilibrium. Thermodynamics, hence, gives us information on the maximum efficiency that is possible in a particular process (i.e. when equilibrium is achieved). Thermodynamics also gives us information on the ideal performance of the separation units in batch as well as continuous mode but cannot tell us anything about the speed of the process. To achieve this ideal performance may require a long time and this may not be practically feasible. The time required to achieve a certain performance is determined by the rate of the process. A particular process may have a desirable equilibrium but the approach to the equilibrium may be slow and vice versa. This is similar to the trade-off seen in the thermodynamic equilibrium and the rate of reaction in the oxidation of sulphur dioxide to sulphur trioxide in the sulphuric acid manufacture based on the contact process seen earlier.

Applications of thermodynamics in the area of chemical reaction engineering are discussed later.

Momentum Transfer

The role of a chemical engineer is to manufacture chemicals on a large scale. When we have a continuous process we need pumps to transport liquids from the storage tank to the reactor or from one unit to another. The engineer must know how to select a pump which he can make the best use of. For example, should he opt for a reciprocating pump or a centrifugal pump? How does one decide the wattage or horsepower rating of a pump? How would you calculate this?

Pumps in a household are used to pump water from the underground storage tank to the overhead water tank. The wattage of the pump depends on several variables, for example, the height to which the fluid has to be raised, the flow rate or velocity of the liquid that you desire, the properties of the liquid such as viscosity and density, nature of the pipe surface (whether it is smooth or rough). This last aspect is determined by the material of construction of a pipe.

Pumps are required even if liquids do not have to be lifted up since they help in overcoming the frictional pressure drop as a fluid flows in a pipe. This pressure drop arises since fluids have viscosity. Viscosity in a fluid acts like friction. This generates a shear or tangential stress which acts in the direction opposite to the motion of the liquid. When the liquid moves in a pipe it loses energy in overcoming this opposing stress. This results in a drop in the pressure which can be viewed as a form of energy loss and the pump helps in overcoming the pressure drop.

Another important aspect in transporting liquids is the choice of the material of construction of the pipe. For instance, we have to ensure that the liquid does not react and corrode the pipe material.

Another area where momentum transfer is useful is in the design of stirrers in reactors. In a reactor two or more reactants are mixed so that they can be in intimate contact. The reaction can occur only when the molecules of the two reactants are close to each other and collide against each other. Hence the objective of the reactor stirrer is to mix the reactants vigorously. For instance, if we consider the nitration of benzene with a mixture of sulphuric acid and nitric acid, then the organic liquid and the aqueous liquids would separate out into two layers if there is no stirring as they are immiscible. The reaction would occur only at the interface where the reactants are in contact and, consequently, would be very slow. To facilitate the reaction between the two liquids the two phase mixture is stirred so that droplets of one phase are dispersed in the other increasing the surface area of contact and thus increasing the rate of product formation.

The engineer has to design or select the stirrer which is most effective. This includes determining the number of paddles, the shape of the paddles, etc. He is also interested in determining the power required for stirring the mixture. This would again depend on the properties of the fluids such as density and viscosity, the speed or revolutions per minute (rpm) of the stirrer, the geometry of the vessel, for example, the aspect ratio (ratio of diameter to height of a cylindrical vessel).

In the earlier days the design of stirrers was based on empirical correlations. These were developed from experimental data. A lack of scientific rigour in this approach resulted in a low level of confidence in these relations. More recently, the approach used is more scientific and based on computational fluid dynamics. This is based on the principles of conservation of mass and momentum. This has a more fundamental basis making it very versatile. It helps us carry out the design of stirrers and relate it to basic physical properties. Since the approach has a scientific basis it is more reliable. We will see applications of this approach and its basis in some detail in Chapter 5 where we discuss scale-up issues.

Heat Transfer

An important contribution to the costs of running a plant comes from the utilities, i.e. power and steam requirements. It would be criminal to waste any heat generated in a unit. This should be recovered to make the process economical. Heat transfer, hence, plays an important role in reducing the cost of running the plant and can decide whether a profit or loss is made in the operation. In a sulphuric acid plant the heat liberated by the exothermicity is used to generate steam and produce electricity by running a turbine.

We now take an example to illustrate where and how heat recovery can be carried out. The stream exiting a reactor, sustaining an exothermic reaction, is at a higher temperature than the feed stream. The energy (enthalpy) present in this stream can be used to preheat the reactants thereby saving costs associated with separate energy requirements for preheating the feed (Figure 3.2). This can be achieved using heat exchangers. A simple design of a heat exchanger is an annular cylinder where one fluid stream flows through the central core and another through the annular region (see Figure 3.3). The two fluids can flow co-currently, i.e. in the same direction or counter-currently, in the opposite direction. In this arrangement the cold fluid entering gains heat and the hot fluid entering loses heat. The chemical engineer must know how the temperature of the streams varies axially. A schematic diagram of this is shown in Figure 3.3 for the co-current and counter current operations.

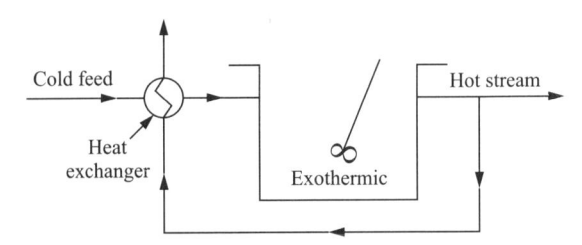

Figure 3.2 Recovery of heat from the exiting hot stream of a reactor to preheat the reactants.

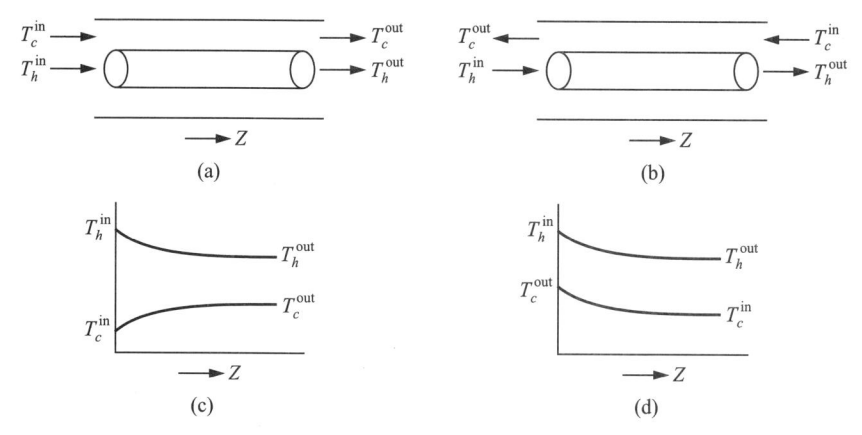

Figure 3.3 Schematic diagram of a co-current (a) and counter current (b) heat exchanger. Temperature profiles along length of co-current (c) and counter current heat exchanger (d).

Let us now see how obtaining the information on the temperature profile is different from energy balance equations which you have solved in your high school classes. Consider a mass m_H of hot water at temperature $T_H^{initial}$ which is added to mass m_C of cold water at temperature $T_C^{initial}$. On mixing these two masses we attain a uniform temperature T^{final}. This can be obtained from an energy balance which states that the energy lost by the hot liquid must equal that gained by the cold fluid.

$$m_H C_p (T_H^{initial} - T^{final}) = m_C C_p (T^{final} - T_C^{in})$$

The above equation can be solved for T^{final}. This is purely based on energy balance. However, the time required to attain this uniform final temperature is not known from this approach, i.e. the information on the rate of heat transfer is missing. All we can determine is the final temperature at equilibrium. This is what thermodynamics tells us.

Similarly, in the heat exchanger problem, the heat lost by the hot fluid must equal to that gained by the cold fluid, under some ideal conditions.

$$\dot{m}_H C_p (T_H^{inlet} - T_H^{out}) = \dot{m}_C C_p (T_C^{out} - T_C^{in})$$

Here the '.' represents the mass flow rate of the two streams. In this problem for a given flow rate combination, one of the temperatures can be determined if three others are specified, from thermodynamics.

However, when an experiment is carried out for a given combination of inlet temperatures, the outlet temperatures are uniquely determined. Hence, it would appear that it is enough to specify only the two inlet temperatures to determine the two exit temperatures. To find the two exit temperatures for a given combination of inlet temperatures information on the rate of heat transfer is required. As a result of the heat transfer, the temperature of the hot stream and cold stream changes. The rate of heat transfer per unit area is proportional to the temperature difference between the two fluids and the proportionality constant is called the *heat transfer coefficient h*.

The energy balance for each stream now is

$$\dot{m}_H C_p \frac{dT_H}{dz} = -h 2\pi R(T_H - T_C)$$

$$\dot{m}_C C_p \frac{dT_C}{dz} = h 2\pi R(T_H - T_C)$$

This will be formally derived in some later courses. These two equations can be solved to get unique profiles only if the inlet conditions at $z = 0$ are known.

At $z = 0$;
$$T_H = T_H^{in}$$
$$T_C = T_C^{in}$$

On integrating the differential equations subject to the above conditions, we obtain T_H, T_C as a function of z. By setting $z = L$, the exit temperature can be found. Using the rate of heat transfer we obtain the temperature profiles along the heat exchanger. Here at every point we use the fact that the rate of heat loss by one fluid is equal to the rate of heat gained by the other fluid. Using the rate of heat transfer we can determine the profiles as well as the exit temperatures when only two of the temperatures are specified.

In a big plant we have to choose from several options for heat recovery. For instance, several streams may be available for heat exchange. The hot and cold streams should be combined for heat exchange such that the heat recovery over the entire plant is maximum. This is an optimization problem.

Critical thickness of insulation

We now take up an example which illustrates why scientific understanding is important to analyse the behaviour of a system. Consider a hot cylinder (pipe with hot contents), i.e. a cylinder at a temperature higher than the ambient. It will lose heat to the surroundings and as a result its temperature would decrease. One way to prevent or slow down this is to add a layer of insulation to the outer wall of the cylinder. One expects the heat loss to the ambience to decrease because of this. However, there could be conditions when the heat loss to the ambience would increase by the addition of the insulating layer. Can you explain how this could happen?

This could arise since when an insulating layer is added, the surface area of contact with air increases. The heat loss tends to increase because the surface area available for heat transfer increases. At the same time the resistance to heat transfer tends to increase due to the increased thickness across which heat has to flow. The increase in resistance arises since the thermal conductivity of the insulator is much lower than that of the conductor. For small thickness, the surface area effect dominates and the rate of heat loss increases. Hence there is a critical thickness which the insulation thickness has to exceed to ensure that the insulation reduces the heat loss. Below this critical thickness, the increase in the surface area dominates the increase in the resistance due to insulation and this causes the heat loss to increase. This example should motivate engineers to be equipped with the ability to physically reason out observations which may be contrary to intuition.

In several processes where heat transfer occurs phase changes are induced. Examples of this are *evaporators* and *condensers*. In the sugar industry, for instance, the water in the solution containing sugar has to be evaporated. *Evaporation* is an energy-intensive process and hence methods to carry out evaporation efficiently have to be designed so that wastage of any heat from the streams leaving the system is minimized. For instance, this leads us to carrying out the evaporation in multiple stages. The challenge here is that the phase change introduces new physics and this has to be incorporated scientifically in designing equipment. The flow, for instance, changes from single-phase flow to two-phase flow. The design of these equipments like boilers, etc. was done empirically in the early days since the science behind the various processes was not clearly understood.

Mass Transfer

Naphthalene balls or moth balls are used extensively to preserve woollen clothes and as a deodorant tablet in toilets in households. As time passes the naphthalene present sublimes and the size of the balls shrink. One question which may occur to us is how long does it take for a moth ball to disappear. This happens because the naphthalene has a vapour pressure of 0.087 mm of Hg at 298 K. When the moth ball is in stagnant air, as in a closed closet, there is a thin layer very close to the surface of ball which contains vapours of naphthalene. The partial pressure in this layer can be assumed to be that of the vapour pressure. The ambient air for the sake of simplicity can be assumed to be infinite in extent. The concentration of naphthalene present can be assumed to be zero in the bulk atmosphere outside the thin layer. We assume here that the concentration in the bulk gas does not change significantly during the entire process. This assumption is valid when the volume of the bulk gas is relatively large. The concentration gradient or the difference in partial pressures of naphthalene close to the surface and in the bulk atmosphere results in a flux of naphthalene into the atmosphere just as a temperature gradient causes a heat flux. In addition to the concentration gradient the diffusivity a molecular property determines the flux of the species. This flux can be used to determine the time taken for a specific reduction in the size of the naphthalene ball.

This problem is analogous to the case of a hot sphere at high temperature in stagnant air at ambient conditions. The sphere will lose heat till the final temperature of the sphere will be the same as the ambient conditions. This information follows from thermodynamics. The time required to cool the ambient conditions, however, is determined by the rate of heat transfer and this cannot be obtained from thermodynamics. The problem of naphthalene ball losing mass has an added degree of complexity in that its size changes as time elapses.

The above example is characterized by mass transfer. The driving force for mass transfer is a concentration difference or for non-ideal systems (in general terms) a difference in chemical potential. A chemical species moves from areas of high chemical potential to areas of low chemical potential. For a single-phase system mass transfer stops when the concentrations become uniform. This is achieved by diffusion in a stagnant system or by mixing using a stirrer. The final state is that of uniform composition in a single-phase system. Consider a glass of water to which you add sugar syrup (a teaspoon of concentrated sugar solution). At the beginning the region containing sugar solution has a high concentration of sugar, while the rest of the solution has a low concentration. If this solution is left to stand and we wait for a sufficiently long time to lapse the concentration of sugar will become uniform in the solution. This is the final state. The state of uniform concentration is achieved through "diffusion". Diffusion occurs whenever a concentration difference exists and it smears out these differences. The analog of diffusion in mass transfer is conduction in heat transfer. In the latter, when there is a temperature gradient in a system, conduction reduces it by transferring heat from the high temperature region to the low temperature region. Thermodynamics gives us information on the final state of equilibrium. However, the time taken to reach this state depends on the rate of mass and heat transfer. This is determined by diffusivity and thermal conductivity, which are properties of the system. If these are high the time taken to reach, equilibrium is low.

Distillation is an example of a two-phase gas liquid system where mass transfer occurs. Consider a mixture of benzene and toluene in the liquid phase. This liquid mixture is taken in a closed vessel. Our interest is in separating the two components present in this mixture. Let us heat this mixture so that the temperature lies between the boiling points of the two components. The vapour phase generated will be rich in the more volatile component (the one with the lower boiling point) benzene which has a boiling point of 353 K. Similarly, the liquid phase will be rich in the less volatile component, toluene which has a boiling point of 384 K. For a temperature which lies in between these boiling points, say, 370 K the vapour-phase composition and the liquid-phase composition in the closed vessel (under batch conditions) is determined by thermodynamic equilibrium. The time taken to reach this equilibrium is decided by the rate of mass transfer. By changing the temperature and pressure in the vessel the compositions of the two phases can be altered and a desired degree of separation or purification can be obtained.

In multiphase systems as in liquid-liquid extraction chemical species will often prefer one phase over the other and reach a uniform chemical potential only when most of the chemical species have been transported into the preferred phase. Thermodynamic equilibrium determines the maximum theoretical performance we can expect of a given mass transfer operation. The actual rate of mass transfer will depend on the flowfield present in the system, the diffusivities, etc. The higher the rate of mass transfer, the quicker is equilibrium reached. Let us illustrate this now with a specific example.

Distillation is not the only technique used for separating homogeneous mixtures where there is a single phase containing different components. Consider, for example, a mixture of two liquids whose boiling points are not significantly different. Then the option of using distillation is ruled out. Extraction is another process which can be used to separate and purify a mixture of two liquids. Consider a mixture of two components: zirconium salt (A) and hafnium salt (B) which are completely miscible with each other in the aqueous phase. This example occurs in the nuclear industry. These compounds have very close boiling points and distillation cannot be used to separate them. A third liquid, tributyl phosphate (C) could be added to this mixture such that A is soluble in C whereas B is not soluble in C. Then we can use C to extract A from a mixture of A and B. Depending on the extraction efficiency the concentration of A remaining in the original solution will reduce. The mixture of A in C can be separated by, say, distillation provided their boiling points is sufficiently different.

Figure 3.4 shows a schematic process of solute transfer from L (heavy) phase to V (light) phase. The solvent tributyl phosphate has more affinity

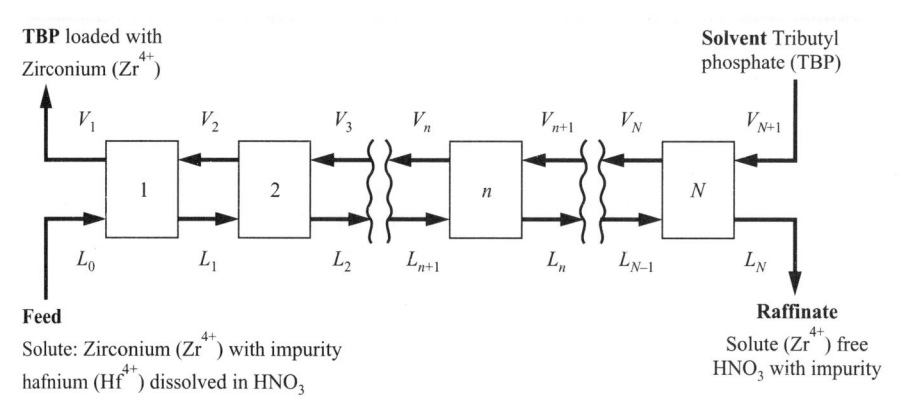

Figure 3.4 A counter-current extraction process to separate zirconium from hafnium with multiple stages.

towards zirconium. When this solvent is contacted with zirconium rich aqueous feed, the solute zirconium transfers to the organic phase, which is termed *extract*. The impurity hafnium gets separated and is discharged as raffinate (waste) stream. The actual process is more complicated and it requires the control of pH and adding a co-solvent to TBP.

Let us consider the extensions to a two-phase system. Here in distillation at equilibrium the vapour (liquid) phase is rich in more (less) volatile component. At equilibrium the concentration in each phase is different. The concentration in each phase remains constant (time invariant) at equilibrium. This is time in variant, but the mole fractions of species in each phase are different. This contradiction is resolved by viewing that, at equilibrium the chemical potential of a species in a phase must equal that in another phase. This chemical potential is a function of composition, temperature and pressure, in each phase. In multiphase systems at equilibrium the concentrations in each phase can be different but their potentials are equal.

In a liquid-liquid extraction system, consider a mixture of water and acetone. The solute is being extracted using chlorobenzene. If the extraction is allowed to proceed for a sufficiently long time the solute will be transferred to chlorobenzene. At equilibrium the concentration of the solute acetone in each phase (water and chlorobenzene) will be different, however the chemical potential of the solute in each phase will be equal. The thermodynamic concept of "chemical potential" is the generalization of concentration in multiphase systems.

For a given objective such as separation there may be more than one possible technique to achieve it. For example, separation of a mixture may be possible by both distillation and extraction. In such a case considerations like energy requirements, costs, etc. are used in arriving at the best approach to carry out the separation.

Another illustration where mass transfer plays an important role and is familiar to students is in drying of objects. It is common knowledge that the rate of drying is more on a dry day than on a humid day. This can be easily understood in terms of mass transfer rates. These rates are determined by the concentration differences of moisture near the object and the bulk atmosphere. On a humid day the concentration difference present between the layer next to the object being dried and the bulk atmosphere is low, as a result of which the time taken to dry is high. This rate can be increased by inducing a flow near the object. This introduces convective mass transfer (due to flow) in addition to diffusive mass transfer (due to Brownian motion).

The application of reverse osmosis, nanofiltration in membrane processes like desalination and haemodialysis seen earlier is based on mass transfer. Here the transport of solutes or solvent occurs across a membrane and is induced by a concentration difference.

Another example where mass transfer plays an important role is in the area of heterogeneous reactions, such as gas-solid catalytic reactions. The active catalyst metal is deposited in pores of an inert carrier. The reactant has to reach the active site inside this pore for the reaction to occur. This occurs primarily by molecular diffusion. As a result of this the concentration at the active site is much lower than that in the bulk gas phase (since diffusion can occur only across a concentration gradient) and this can result in a lower reaction rate than what is expected. This has to be factored in when designing gas solid catalytic reactors otherwise the performance of the actual reactor will be far inferior to the expected behaviour.

Transport phenomena provided a scientific basis to understand the transport process. This has helped improve confidence in design and performance prediction of devices where heat, mass and momentum transfer are important. The concept of *Transport Phenomena* as introduced by Bird, Stewart and Lightfoot brings forward the similarity in the transport of mass, momentum and energy and provides a generalized framework for analysing the system behaviour. These transport processes are analogous. Thus, diffusion in mass transport and conduction in heat transport are similar. In the former this results in a mass flux across concentration gradients whereas in the latter the heat flux is induced across temperature gradients. *Fourier's law of heat conduction*, *Fick's law of diffusion* and *Newton's law of viscosity* established a linear relationship between fluxes and corresponding gradients. The analogous nature of these concepts can be seen in Table 3.1. Similarly, convective transport is induced by bulk flow and this again has the same form for mass, momentum and energy. This approach of analysis provides a better insight into the behaviour of systems as it allows a rigorous scientific basis which includes physical interactions at the molecular level. This was missing earlier where the emphasis was primarily on empiricism.

Table 3.1 Analogy between fluxes, potential gradients and resistances

Ohm's law	$I = \dfrac{\Delta V}{R}$	ΔV potential difference
		R resistance
		I current
Fourier's law	$q = -kA\dfrac{\Delta T}{\Delta X}$	q heat loss, cal/s
		ΔT temperature difference,
		$\dfrac{\Delta X}{kA}$ thermal resistance
Fick's law	$N_A = -DA\dfrac{\Delta C}{\Delta X}$	N_A moles transferred, moles/s
		ΔC concentration difference
		$\dfrac{\Delta X}{DA}$ diffusive resistance
Newtonian fluid	$F = -\mu A\dfrac{\Delta V}{\Delta y}$	F rate of change of momentum
		ΔV velocity difference
		$\dfrac{\Delta y}{\mu A}$ viscous resistance

The approach using transport phenomena led to a significant increase in the role of mathematics in chemical engineering.

Chemical Reaction Engineering

In Chemical Reaction Engineering (CRE) the concepts of thermodynamics and physical chemistry are integrated to help in designing a reactor. The chemist is successful in establishing the conditions for a reaction on a small scale. He would be interested in understanding the mechanism of the reaction—whether there is an electrophilic substitution or a nucleophilic substitution. The engineer is usually not concerned with these details. He is interested only in using information from thermodynamics (Gibbs free energy change and heats of reactions, for instance) and physical chemistry (reaction rates) to help design reactors. He is interested in designing a reactor (determining the volume) for a particular production rate or to predict the performance of the reactor for a given design.

The free energy change of a reaction indicates whether a reaction is feasible or not. If this were negative, then the reaction is feasible. However, this does not give us any information on how fast the reaction is. This latter information is contained in the kinetics of the reaction.

For instance, if the reaction rate is very slow the reactants have to spend sufficient time in the reactor for a significant amount of conversion to products. This would mean for a given volume of a reactor having a low flow rate of the reactants or for a given flow rate (residence time) having a high volume of the reactor. The time spent by reactants in the reactor on an average (residence time) is given by reactor volume/flow rate. For slow reactions in order to have enough residence time, the volume of the reactor should be high and the flow rate of the reactants should be low. For fast reactions the amount of residence time required is less and, hence, we can have small reactors through which reactants flow at high rates translating to higher rates of production.

In addition to the reaction rate we need to use information about the heats of reaction which comes from thermodynamics. If a reaction is endothermic as the reaction progresses the temperature would drop and the reaction would be slowed down (recall the Arrhenius temperature dependency of reaction rates). This decrease in temperature would hence have to be negated and this can be achieved by supplying heat to the reactor. The amount of heat that has to be supplied to carry out an endothermic reaction at a constant temperature has to be calculated from the energy balance.

Analogously, for the case of exothermic reactions we need to remove heat to prevent the temperature from rising drastically. The exothermic reaction is an autocatalytic system or one which is self-propagating. As the reaction progresses heat is liberated and the temperature rises. As the temperature rises the reaction rate increases. This causes a further increase in the rate of heat generation. This in turn causes the temperature to rise even faster and further increases the heat generated. Thus, we have a positive feedback effect and this can result in what is called a "runaway condition". This can give rise to an unstable steady state in a continuous reactor. A rapid drastic increase in temperature would result in possible boiling of the reactant mixture and can cause an increase in pressure in the reactor and an accident. Hence reactors which sustain such exothermic reactions have to be provided with cooling systems. Two simple mechanisms of providing heat to these reactor systems are jackets surrounding the reactor or coils inside the reactor (Figure 3.5). Through the jacket or the coil another liquid usually a coolant flows when an exothermic reaction occurs. The rate of heat transfer depends on several factors such as the temperature difference between the fluids, flow rate of coolant, and thermal conductivity of the wall. This example shows that the concepts in different subjects (heat transfer and chemical reaction engineering) are interlinked.

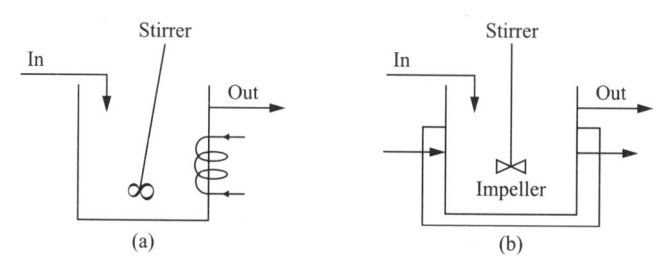

Figure 3.5 CSTR with coolant flowing through (a) cooling coil, and (b) jacket.

The emphasis on thermodynamics and physical chemistry helps represent the behaviour of a reaction at an abstract level. This helps the engineer talk about the behaviour of reacting systems in terms of their characteristics like fast reactions, exothermicity, etc.

The reactions occurring are classified as homogeneous or heterogeneous depending on the number of phases participating in the reaction. The classical homogeneous reactors are:

(a) Batch reactor
(b) Continuous stirred tank reactor (CSTR)
(c) Plug flow reactor (PFR)

We will now see the characteristics of these reactors under ideal conditions. In the batch reactor the reactants are added to the reactor and the contents are well mixed. This serves to homogenize the conditions in the reactor, i.e. maintain uniform concentration and temperature. The conditions of temperature, pH, etc. are maintained in the reactor for the reaction to occur. At the end of the reaction time the reactor contents are emptied.

CSTR is an example of a continuously operated reactor. This can be viewed as a tank which has an inlet and an outlet. The reactants are fed into the reactor through the inlet while the products and uncoverted reactants exit through the outlet. The conditions for the reaction are maintained inside the reactor and it is assumed that there is no reaction occurring in the inlet and outlet pipes. The composition, temperature, etc. in the exit pipe and inside the reactor are assumed to be the same. There is a sudden drop in the concentration and temperature from inlet to reactor. The composition inside the reactor is assumed to be uniform via stirring.

PFR can be viewed as a tubular reactor where the concentration and temperature vary along the length. Reactants are fed in at one end of the tube the inlet. The velocity is assumed to be uniform across the cross-section for an ideal PFR. The concentration and temperature are also assumed to be radially uniform. The products are taken out of the stream leaving the exit of the pipe.

The performance of the above reactors for different reactions has to be predicted. This is done under the ideal conditions or assumptions stated above, i.e. of being well mixed, etc. There can be situations when these assumptions are not valid and this would cause deviations in the reactor performance. The engineer has to determine what is the cause of this deviation by performing certain diagnostic tests and bring the reactor behaviour to as close to ideal as possible.

Chemical Reaction Engineering (CRE) is concerned with

(i) Determining the volume of a given reactor and predicting its performance.
(ii) Determining the operating conditions such as feed temperature, composition, and flow rate for safe operation.
(iii) Determining if the reactor designed and being operated is ideal.

In several situations multiple reactions occur with a set of reactants. Here, in addition to the desired product being formed, undesired products are also produced. For instance, the product may form isomers and only one isomer may be of interest. Chemical Reaction Engineering deals with determining the operating conditions to maximize the production of desired product. This could be achieved by changing the operating conditions, i.e. composition or the feeding policy of different reactants. An example of a series parallel network of reactions of industrial importance is the formation of maleic anhydride from butane oxidation. This can be written as

$$\text{Butane} + \text{Oxygen} \rightarrow \text{Maleic anhydride}$$

$$\text{Butane} + \text{Oxygen} \rightarrow \text{Carbon dioxide}$$

$$\text{Maleic Anhydride} \rightarrow \text{Carbondioxide}$$

Conditions in the reactor have to be chosen to ensure that the production of maleic anhydride the desired product is maximized.

Another important class of reactions is the heterogeneous catalytic reaction. These reactions are carried out in packed beds. Catalyst particles are porous and the active sites where the reactions occur are in the pores of a substrate. Here to study the reaction, the transport of the reactants to the catalyst site from the bulk gas phase, the formation of products at the site and the transport of products from the site to the bulk phase has to be understood. Several gas phase reactions are catalysed by expensive metals such as Pt, Au, Rh. Catalyst particles are not made completely of the active metal. The reaction takes place on the surface of the catalyst and it is desirable to have as large a surface area per unit volume. One way to achieve this is to have small particles or powders.

The disadvantage here is the pressure drop across this bed of particles would be very high. To overcome this an inert material like silica SiO_2 or alumina Al_2O_3 is taken and it forms a porous substrate as carrier. The noble metal like Pt is deposited in the pores of the substrate. The reaction occurs at the sites of the catalyst inside the pores (Figure 3.6). The reactant gas flows around the particle. The reactant has to reach the active site for the reaction to occur. The cylindrical pores in the catalyst are very fine and here no flow or bulk transport can occur. The reactant (product) is transported by diffusion from the bulk to the active site. As a result of this the concentration of the reactant at the active site is much lower than that in the bulk phase (since diffusion needs a concentration gradient). Consequently, the reaction rate will be much lower than what is anticipated. This has to be accounted for in the calculations. These examples show that mass transfer plays a vital role whenever there are multiple phases present.

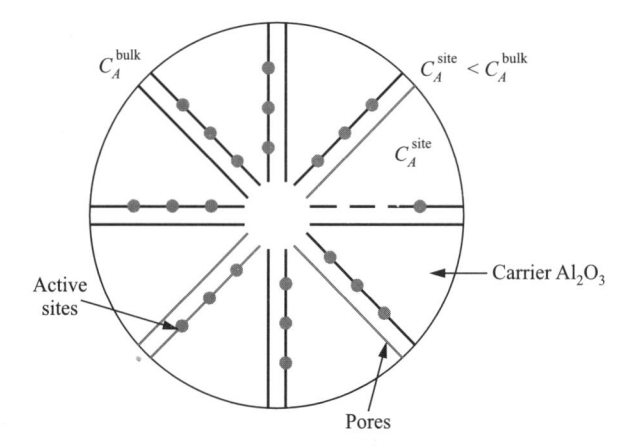

Figure 3.6 Porous catalyst particle.

Remarks

It is easy to see that the different rate processes (momentum, heat and mass) are coupled with each other. For instance, in distillation we would have to supply heat to ensure separation. Here heat and mass transfer occur simultaneously. As far as the B.Tech programme is concerned, the various effects are treated as being independent. For instance, the heat required for a distillation column can be calculated but in a course in mass transfer we will not focus on how the heat will be supplied but assume the heat is available for affecting the separation. In the course on chemical reaction engineering again the focus is on the reactor design and not how the heat is supplied or removed. In other words, the way the courses are structured is such that each course will concentrate only on one aspect at a time. Having mastered the fundamentals of each course, the student can then integrate the concepts to design plants in a holistic way.

4

Role and Importance of Basic Sciences in Engineering

Introduction

The common misconception which prevails is that chemical engineering is based primarily on chemistry. It is not true. In fact, chemical engineering involves a combination of principles of mathematics, physics, chemistry, biology, and even economics. Biology comes in since some processes such as fermentation involve microorganisms and are biochemical in nature. Economics is necessary so that the cost of production can be addressed. This ensures that a technological concept is also economically viable.

All systems must satisfy the principles of conservation of mass, momentum, and energy. These principles are used to solve problems concerning the design of an equipment (such as a reactor or a distillation column) for a desired performance level or predicting performance for a specific design. The former would answer: what should be the size of a reactor for a given production rate? The latter would answer the question given the size of a reactor what would be the production rate. It would be no exaggeration to say that the role of physics and mathematics dominates the role of chemistry in traditional chemical engineering.

In this chapter, examples of the interaction between these disciplines will be discussed as they arise in different applications. First, we will see different examples wherein we use physics and mathematics to analyse a situation. Physics is used to formulate equations and mathematics is used to solve

74   *Introduction to Chemical Engineering*

equations. Equations arise from the applications of the fundamental laws which any system must conform to. The three fundamental laws all systems must obey are the *laws of conservation of mass*, *momentum*, and *energy*. The student must have come across these concepts in different contexts in his or her school education. One of the main responsibilities of the chemical engineer is to understand how to apply these laws to specific systems and problems and obtain useful information from them. Here we restrict ourselves to simple ideal cases and explain some of the issues which arise in analysing these systems.

In the first half of this chapter, we focus our attention on closed systems. In the second half, we will derive the fundamental law which governs the behaviour of open systems. This will result in an equation which is very important in chemical engineering. This equation will arise in different forms in different situations or applications. It is used to analyse the behaviour of different systems in various contexts. Applications of this equation in different situations give rise to different kinds of mathematical equations: algebraic equations, ordinary differential equations, and partial differential equations. In this chapter we will see different contexts in which these equations may arise. This will help the student realize the relevance of various courses in mathematics to engineering. When pursuing the courses in mathematics he would realize that there are engineering applications to the mathematical tools he is learning.

An Application of Conservation of Mass in a Closed System

To begin with we start with an example based on the principle of conservation of mass, in a nonreacting system. This example gives rise to a system of linear algebraic equations and helps us understand the relevance of linear algebra in chemical engineering. This is an example of an application to an isolated system or a closed system.

The two compounds sodium chromate Na_2CrO_4 and sodium dichromate $Na_2Cr_2O_7$ contain the same three elements: sodium, chromium, and oxygen. A mixture of these two compounds, containing x gram of the chromate mixed with y gram of the dichromate will also contain the same three elements. Given such a mixture, it is of interest to determine the masses of the two compounds present in the mixture, i.e. x and y.

How can we go about doing this? What information is required to determine x and y? What experiments can be thought of to address this question?

The mixture can be weighed to determine the mass. The mass of the mixture m_{total} can be found. This gives us a relation of the form

$$x + y = m_{total} \qquad (4.1)$$

This gives us only one equation while we have to determine two unknowns x and y. It is, hence, necessary to determine at least one more relationship to help determine x and y.

This relationship can be obtained from the information on the amount of individual elements Na, Cr, O present in the mixture, i.e. a species or component balance. Let the molecular mass of the two compounds chromate and dichromate be M_c and M_d respectively (this can, of course, be calculated from the formula and the atomic mass). From the molecular formula the fraction of the mass being contributed by an element to the molecular mass can be determined.

To obtain the additional equations for x and y the mixture is analysed for the mass in grams of the different elements in the mixture through an analytical method. This could be based on a spectroscopic technique like inductively coupled plasma-optical emission spectroscopy (ICP-OES) or an atomic absorption spectroscopy (AAS). The mass in grams of an individual element in the mixture is obtained experimentally using these techniques. For sodium let it be denoted by M_{Na}. This information can be used to write a mass balance for each species.

From the formula for each of the compounds the mass balance for various elements can be written. For sodium, this is given by

$$\frac{2 \times 23 \times X}{M_c} + \frac{2 \times 23 \times Y}{M_d} = M_{Na} \qquad (4.2a)$$

Similarly, for Cr the mass balance is

$$\frac{1 \times 52 \times X}{M_c} + \frac{2 \times 52 \times Y}{M_d} = M_{Cr} \qquad (4.2b)$$

And for oxygen, it is

$$\frac{4 \times 16 \times X}{M_c} + \frac{7 \times 16 \times Y}{M_d} = M_o \qquad (4.2c)$$

At first glance it would appear that we now have four equations to determine the two variables x and y. Do we really have four equations?

Are the four equations independent? Adding the three Eq. [4.2(a–c)] results in recovering Eq. (4.1). Hence, the four equations are not independent. It is clear that we have only three species or component balance Eq. [4.2(a–c)] or the overall mass balance and two of the three component balance equations from the set [4.2(a–c)] which form an independent set. This, however, does not solve our problem since the number of variables to be found is two (equal to the number of compounds in the mixture) but we have three equations (corresponding to the number of species or elements in the compounds).

How do we find two variables x and y which satisfy all three equations? Do we choose only two of the equations and if so how do we decide which two? In the absence of any experimental error [when we are able to determine the right-hand sides of Eqs. (4.1 and 4.2) accurately to infinite precision] if we choose two of the three equations and find a solution it will be found that the solution satisfies the third equation as well. The system of three equations in this case constitutes a "consistent" system of equations.

In a general situation two of the equations can be chosen and a solution can be found but the solution may not satisfy the third equation. This is likely to be the case since the right-hand sides are determined experimentally and are prone to experimental error. This kind of a system of equations is called an "inconsistent" set. In such a situation, how can we go about finding a solution to the system? Do we choose two equations out of the three randomly? Or do we use information from all the three equations?

Here is a way out. If errors in experimental measurement of some of the species (say, that of chromium) is more than that of the other we can give less importance to the chromium measurement or even neglect it and consider only the masses of the other species obtained. In this case it would be improper to find a solution which satisfies Eq. [4.2(b)]. It is then justified to consider only Eqs. [4.2(a) and 4.2(c)] to find the solution.

If the accuracy with which the three masses are determined experimentally is equal, then we must consider all the three equations with equal importance. Is it desirable to use all the measurements if all of them can be measured to the same level of accuracy? The answer is *yes*. There is redundancy of information when all measurements are used and it is best to use the redundant information to arrive at the "best" possible solution. By the best possible solution, we mean a solution which satisfies all the equations with a minimum error. Such a situation (where there are more linear algebraic equations than the unknowns) occurs often in chemical engineering and we will now see how such a system of equations can be solved.

Equations of the above kind are called *linear algebraic equations* since they are algebraic (they do not contain any derivatives and hence are not differential equations) and they are linear in the variables to be found x, y. By linear we mean the equations that contain the dependent variables x, y to the first power. Such equations can be written or recast in the vectorial form

$$Ax = b \qquad (4.3)$$

$$X = \begin{bmatrix} x \\ y \end{bmatrix} \quad A = \begin{bmatrix} \dfrac{2 \times 23}{M_c} & \dfrac{2 \times 23}{M_d} \\ \dfrac{52}{M_c} & \dfrac{2 \times 52}{M_d} \\ \dfrac{4 \times 16}{M_c} & \dfrac{7 \times 16}{M_d} \end{bmatrix} \quad b = \begin{bmatrix} M_{Na} \\ M_{Cr} \\ M_O \end{bmatrix}$$

Here A is an ($m \times n$) matrix which is not square (m is not equal to n). Here x is a vector with n elements and b is a vector with m elements. This is a system of m equations in n unknowns. Here n represents the number of compounds (in our case two) contributing to the mixture while m is the number of chemical elements (in our case three) present in the mixture. The system of Eq. (4.2) is written in a compact vectorial form [Eq. (4.3)].

If the number of equations m is less than the number of variables n, then the system can have an infinite number of solutions. This is easy to see as we can arbitrarily assign a value to $(n - m)$ variables and solve for the remaining m variables. Since the variables can be assigned arbitrarily and chosen arbitrarily an infinite number of solutions can arise.

For the case when $m = n$, a unique solution exists if the determinant of A is non-zero, i.e. the rows and columns of A are independent.

If the number of independent equations m is more than the number of variables n, the solutions which are "best" representative of the system, in some sense, have to be found. This is a situation we are interested in.

Systems where the number of equations is more than the number of variables are solved by multiplying both sides of the equation on the left with A^t the transpose of the matrix A. (The transpose is obtained by interchanging the rows and columns of A.) This gives rise to a square system (where the number of equations is equal to the number of unknowns) since $A^t A$ is an $n \times n$ matrix which can be inverted. The equation can be transformed as

$$A^t A x = A^t b \tag{4.4}$$

This system can be solved by matrix inversion using the standard techniques to obtain

$$x = (A^t A)^{-1}(A^t b) \tag{4.5}$$

Standard methods can now be used to solve this since the resulting system of equations is such that the number of equations is equal to the number of unknowns. When the modified set of equations above is solved a linear combination of the original equations is solved such that the number of equations is equal to the number of unknowns. While there are several ways to take a linear combination of the original equations the combination generated by the above method (by premultiplying with A transpose) gives the "best" possible solution. The solutions obtained by this method are optimal or best in the sense that the least squares error in the different equations is minimized when this solution is used. It will be formally shown later through an example that this solution indeed minimizes the least squares error.

The two compounds, sodium chromate and sodium dichromate, have different molecular formulas. The composition (mass fraction or mass per cent) of the various elements in the individual compounds is different. It is clear that this fact is exploited and used and forms the basis in the above approach to determine the masses (x, y) which have been mixed.

A Practical Application of this Idea to the Source Apportionment Problem in Air Pollution

The idea discussed just now forms the basis on which the source apportionment is carried out in an air pollution study. Here the objective is to determine the

78 *Introduction to Chemical Engineering*

different quantitative contributions of various sources of pollution to the pollution levels at a point. In any region there are likely to be multiple sources contributing to the pollution levels. In particular the quantitative contribution of various sources has to be determined. It is desirable to have a quantitative estimate of the contributions of various sources. This will help us come up with policy decisions to control the pollution levels. For instance, if it is found that vehicles are contributing significantly to the pollution levels, then it is necessary to have a control policy to implement tighter emission norms for vehicles. If a neighbouring factory is found to be contributing to the pollution levels, then the emissions of the factory have to be controlled. The problem of determining source contributions is also called *receptor modeling* and is based on the concept of *chemical mass balance* (CMB). This has been used extensively in atmospheric pollution studies. Similarly, if there are several industries which are polluting a water stream, then the contributions of each industry can be found if the composition of the effluent from each industry is known. The principle on which CMB works is the same as the one used to solve the sodium chromate and dichromate problem.

One of the important parameters which determines the pollution levels of ambient air is PM_{10}. This consists of *particulate matter* whose size is less than 10 microns. This particular size fraction is important since it has detrimental effects on health. It is desirable to find out which of the sources are contributing to the levels of PM_{10} at a point where it is measured. This can be estimated by collecting the dust samples at a monitoring station.

The dust sample consists of particles which come from various sources. Each source emits dust particles with its own unique composition. Thus, the particles coming from vehicles are rich in carbon content while those coming from the roadside dust will be high in silica and other metals. The composition of the dust from each source is called a *source profile*. This source profile is similar to the composition of a particular compound which is determined when the molecular formula is known. The dust sample collected at a site is hence a composite mixture of particles with different compositions. Our objective is to determine the contributions coming from the various sources. One can see the similarity between this problem and the problem of finding the composition of the mixture of sodium chromate and dichromate discussed earlier.

The analysis of the chromate problem gives us a hint on how to proceed here. The dust sample collected has to be analysed for various elements or species present, i.e. it has to be speciated. The idea here is to determine the species concentrations of various elements present in the dust particles collected. This is done experimentally and forms the right-hand sides of Eq. (4.2). This mixture is composed of dust particles coming from various relevant sources prevailing in a region such as paved road dust, factory emissions, and vehicular exhaust. The dust coming from each source has a specific composition and this is represented in the form of a source profile (similar to the mass fraction of various elements in the dichromate and chromate as determined by the molecular formula). The source profile consists of the composition of elements in the particulate matter coming from a specific pollution source.

Table 4.1 Experimental measurements in verification of Ohm's law

Current (I)	1	2	3	4	5	6	7	8	9	10
Voltage (V)	15.0	23.0	36.0	39.0	54.5	58.0	73.9	83.0	87.0	105.0

The columns of the matrix A in Eq. (4.3) now consists of the source profiles of various sources. A mass balance for each of the elements gives rise to a system of linear equations. To obtain a physical solution the number of elements analysed in the collected dust must be equal to or more than the number of sources whose contributions are being measured. In this problem again a system of linear equations is obtained which can be cast in matrix form as

$$PS = C$$

Here P represents the matrix of size ($sp \times so$) of source profiles. The columns of P represent the source profiles of the different so sources. S ($so \times 1$) is the column vector containing the contributions of the sources which have to be found and C is the vector ($sp \times 1$) containing the composition of the elements which are analysed in the dust gathered.

This is a simple example of how *conservation of mass* can help us get useful information. Here we see the interplay of mathematics (techniques used to solve the equations) and chemistry (experimental techniques which are used to analyse the compositions) and physics since the equations come from the conservation of mass.

Here only the basic methodology and the idea behind the strategy to solve the system of equations have been explained. In reality the solution methodology is more involved and takes into account the uncertainty in the concentration measurements (C) and source profiles (P) as well. Besides, the solutions S which represent source contributions have to be positive to be physically meaningful. We can ensure this when the system of equations is solved by imposing a constraint to ensure that the solutions are positive.

Another area where a similar situation arises (the number of equations being greater than the number of unknowns) and the above method of solution is used is the *method of least squares*. Let us illustrate this with an example which you are familiar with. The relationship between two variables in several systems is linear. For instance, the voltage across a resistor is linearly related to the current flowing through it and the proportionality constant is the resistance of the wire. In several cases the relationship between two quantities may not be linear but when we take a transformation of a variable the transformed variable may have a linear relationship with the other variable. For instance, the activity of a radioactive substance varies exponentially with time.

The logarithm of the activity, however, depends linearly on time or the rate of change of activity depends linearly on the activity level.

$$A = A_0 \exp(-kt)$$

This shows that the activity decreases exponentially with time, i.e. in a nonlinear manner. However, the rate of change of activity varies linearly with activity.

$$dA/dt = -kA$$

When two variables are related linearly, the relationship can be represented graphically in the form of a straight line. Frequently due to experimental errors the data points do not lie exactly on a straight line. In such cases the question arises as to what is the best possible fit or which line gives the "best possible fit" of the data. Here by "best possible fit" we mean one for which the error estimated has the least value.

A classical high school experiment in physics is one in which you have to determine the resistance of a wire. This experiment is carried out by varying the voltage across the wire and measuring the current assuming that the resistance follows Ohm's law. The experiment generates a set of current values for a set of voltage values. Thus, if ten measurements are made, ten values each for current and voltage are obtained. The resistance of the wire is determined by finding the ratio between V and I. Thus, ten values for the resistance are found. These would not all be identical since there is a finite precision with which measurements are made and there are experimental errors. Hence, in a realistic situation different values for the resistance are obtained for each measurement. Depending on the accuracy of the measurement these would be scattered around a mean value. One approach is to find the mean value of these ten measurements and report that as the resistance of the wire. This is possibly what you have done in your high school class.

Is there an alternative way to do this? Is there a better way of doing this?

Let us now discuss the application of the methodology of the source apportionment problem to the example involving the determination of the resistance R of a wire from multiple measurements. Consider the data of voltage and current given in Table 4.1 and assume it has come from experimental measurements.

In this data set the values of the voltage are experimentally found for fixed values of current. The voltage reading can contain errors. Our objective is to determine the resistance using all the k measurements. If the resistance value is R, then the predicted voltage value for the kth data point is $I_k R$. Both V_k and I_k are experimentally measured and known.

Each data point satisfies an equation of the form

$$V_k = RI_k \tag{4.6}$$

Here again, we have the same situation as earlier, i.e. the number of equations is more than the number of variables. The number of equations which has to be satisfied is equal to the number of data points (ten in this case). This is more than the number of variables (in this case it is one, i.e. R) to be determined.

The above equation can be written as

$$I_k R = V_k \qquad (4.7)$$

To determine R, we construct the vector of current I (10×1) and the vector of the voltage V (10×1) and write all the equations in a compact form using a matrix notation. The best possible solution is determined as

$$R = (I^t I)^{-1} (I^t V) \qquad (4.8)$$

Using this value of R we can determine the predicted values of V. These are denoted as V_{pred}.

We now show that this value of R minimizes the sum of the squared error $(V_{\text{pred}} - V)^2$ over all the data points. The value of R predicted from the above equation is the one which gives the minimum error.

To prove this we seek R such that we minimize the sum of the errors squared.

$$\text{Minimize} \sum_{k=1}^{N} (V_k^{\text{expt}} - V_k^{\text{pred}})^2 = \sum_{k=1}^{N} (V_k^{\text{expt}} - I_k R)^2$$

Differentiating the above equation with respective R and setting it equal to zero yields

$$2 \sum_{k=1}^{N} (V_k^{\text{expt}} - I_k R) I_k = 0$$

Solving this for the resistance R, we obtain

$$R = \frac{\sum_{k=1}^{N} V_k^{\text{expt}} I_k}{\sum_{k=1}^{N} I_k^2} = \frac{I_k^t V_k^{\text{expt}}}{I_k^t I_k} \qquad (4.9)$$

This is the same as the expression in Eq. (4.8).

Imagine the experimental data, i.e. the values of voltage (on the y-axis) and the corresponding current (on the x-axis) are plotted on a graph sheet. This would be a set of points and our objective is to determine the straight line (since the V–I relationship is linear according to Ohm's law) which best fits all the data points. We would get ten different values of resistance from each experimental data point. This can be averaged and a mean representative value can be determined as discussed earlier. For the data set above let us denote this by R^{mean}. The linear relationship between V and I assuming this R is shown as a solid line in Figure 4.1. It can be seen that most of the data points lie on one side of this line. Is it possible to get a best fit line? By best fit we mean one which minimizes the error between all the experimental measurements of V and the predicted values of V for a given current I using a unique value of R (when $V = IR$ is used). This is obtained using Eq. (4.9) and is shown by a dashed line. Here the experimental points are scattered on both sides of this line.

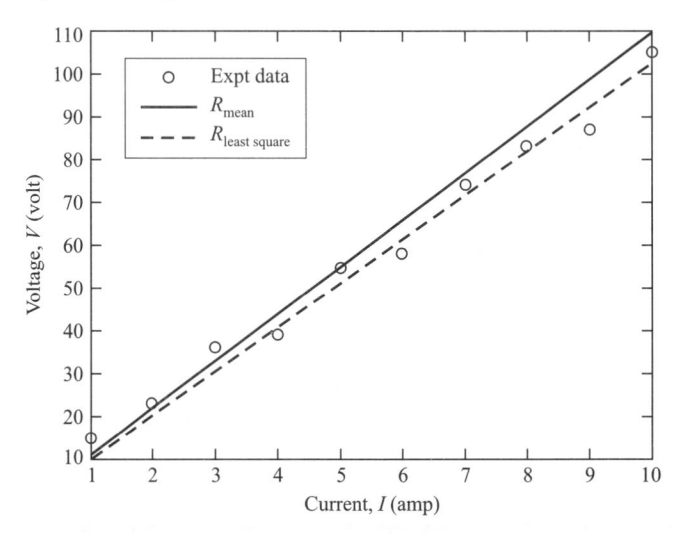

Figure 4.1 Comparison of least squares error prediction with the prediction obtained using the mean value of R from each data point in verification of Ohm's law.

Figure 4.1 shows the two predictions using the values of R obtained using the two methods. One is found by taking the average of individual values of R from each experiment (solid line) and the other by using the least squares technique (dashed line). The dashed line minimizes the error, i.e. length of the vertical lines drawn from the data points to the best fit line. It is seen that the one obtained using the least squares technique passes through the experimental data points more closely. A MATLAB program which generates Figure 4.1 is given in MATPROG 1 in the Appendix. Best-fit lines in packages like MS Excel are based on this principle.

We, thus, see that the approach proposed to solve a system of equations where the number of equations is more than the number of unknowns is such that it is the "best possible solution" in that it mimimizes the error.

Conservation Laws, Closed Systems and Open Systems

The earlier examples illustrate how the concepts from physics or chemistry (conservation of mass for a closed system) can be combined with mathematics to help analyse a practical system.

Let us now quickly recall the different laws of conservation that you have encountered in different contexts in the courses of science in your high school. These are the laws of conservation of mass, momentum, and energy.

The conservation of mass statement that most of you are used to is "mass cannot be created or destroyed". An alternative statement is "Mass of a material body is a constant". Here we are focusing our attention on a body of fixed mass or a fixed collection of molecules. Consequently, for a material body by definition the mass cannot change with time and this is written as $DM/Dt = 0$, where the capital D is used to signify that we are looking at a material derivative, i.e. a fixed collection of particles or a fixed mass. This is true as long as we exclude nuclear reactions, since in nuclear reactions mass can be converted to energy.

The law of conservation of momentum which is taught in high school is *Newton's second law of motion* which states that the net external force acting on a body is equal to the rate of change of momentum of the body.

$$F_{ext} = \text{Rate of change of momentum} \tag{4.10}$$

The law of conservation of energy stems from the *first law of thermodynamics*. It states that the change in internal energy of a system is given by the difference in the heat gained and work done by the system.

$$\Delta U = Q - W \tag{4.11}$$

These laws in the form in which we have expressed them are valid for constant mass systems. Systems are classified as follows.

Isolated systems cannot exchange mass or energy with the environment.

Closed systems cannot exchange mass but can exchange energy with the environment.

Open systems can exchange mass and energy with the environment.

The mass present in a closed system is constant, i.e. does not change with time. The laws of conservation and the forms in which you have come across them in your high school are applicable for closed systems.

However, in most chemical engineering problems we do not encounter closed systems or systems with constant material mass. In the continuous operation of plants we have streams of fluid moving from one equipment to another. The material mass on which we focus occupies different volumes or regions at different instants. These streams could get split into two sub-streams and hence a collection of matter could get broken into two or more parts. Besides, after some time the fixed collection of mass on which the conservation law is applied may have left the system. Applying the conservation principles to a fixed collection of particles or a fixed material mass is not practical in chemical engineering.

Here it is of interest to apply the laws to fixed equipment. This could be a reactor or a heat exchanger or an extraction unit. In this era of continuous processing, streams of fluids flow across these units. For instance, in a continuously operated reactor a stream containing fresh reactant flows in and a partially converted stream containing products leaves the reactor. The reactor hence holds different material masses at different instants since mass comes in and leaves continuously. This is true even in the absence of reactions.

The conservation laws as we know them, cannot be applied to these open systems. They have to be modified as the systems are now open systems. Here the system contains different material mass at different instants. Hence as chemical engineers it is necessary to learn how to extend or modify these conservation laws to open systems which can exchange mass with the surroundings.

In applying conservation laws to chemical engineering systems/units we must know how to apply them to systems through which the flow occurs especially when we are dealing with continuous processes. Our objective is to apply these conservation laws to various units so that we can quantitatively predict their performance or design them. In the first case predicting performance of a unit, say, a reactor, would mean determining how much conversion is obtained in a reactor for a given set of operating conditions and for a given reactor (size) design. In the second case designing a unit, say, a reactor would mean we ask the question what should be the size of the reactor for a desired conversion, i.e. we solve the inverse problem.

The general form of the conservation law for an open system will now be obtained. A formal derivation is not the aim of this chapter. We intend to derive the equation in a general form which the student will come across in subsequent courses. There he will apply the general form of the equation to specific systems.

Before embarking on the extension of the conservation law to an open system a few definitions are given. The focus now shifts to applying these conservation laws to fixed regions in space which can contain different amounts of material mass or different composition of material mass or molecules. For this, we first introduce the concept of a *control volume* (CV) which is a fixed region in space with a defined boundary. This can contain different compositions of material mass since mass can flow across the boundaries of the control volume. The boundary of the control volume is called the *control surface* (CS). A control mass, on the other hand, is a fixed collection of material molecules which can occupy different regions in space (whose volume and size can change).

As discussed earlier from a chemical engineer's perspective, applying the conservation laws to a control mass is not relevant. The control mass chosen moves in space and after some time may have left the plant through the wastewater stream. Alternatively, the material mass may get split into two different streams which flow through two different pipes and it is not possible to keep track of them as one single entity. Hence, we have to understand how to adapt or modify these conservation laws for a control volume or for a fixed region in space which will be occupied by different material mass at different instants. Figure 4.2 illustrates a control volume fixed in space. The flow occurs across the control surface as shown there.

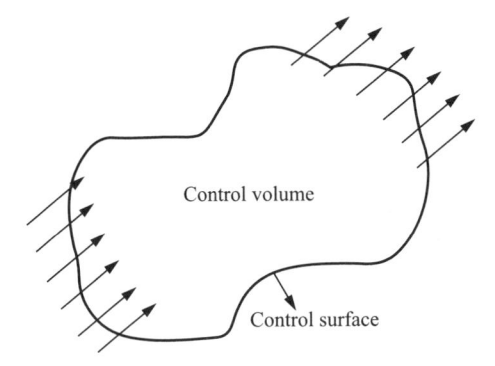

Role and Importance of Basic Sciences in Engineering 85

Figure 4.2 Illustration of control volume (CV) and control surface (CS). Arrows indicate flow across CS.

Infinitesimal Control Volume

We first explain the difference in the two approaches (closed and open systems) using the concepts which you have perhaps come across in courses in calculus. Physically, we apply this concept to an infinitesimal (very small) mass which occupies an infinitesimal volume or a point at any instant. Consider a liquid stream moving in a pipe. This system can be described in one of two ways: (i) focusing our attention at a point in a pipe and recording how a property changes at the point which is occupied by different molecules, and (ii) focusing our attention on a collection of molecules which move so that these molecules occupy different spatial positions at different instants. If we are looking at temperature as the property, then in the former approach we measure the temperature at different instants at a fixed point in space. The rate of change of temperature then is given by the partial derivative of temperature T with time, i.e. $\partial T/\partial t = 0$, at a fixed point. In the latter case when we measure the change in property of an infinitesimal mass or a collection of molecules the change in temperature measured is the total derivative which takes into account the change in temperature due to change in time as well as the change in spatial position. Here the spatial position also changes with time since the mass has moved along with the fluid. The temperature function can be written explicitly as

$$T = T[x(t), y(t), z(t), t] \tag{4.12}$$

Here we explicitly show that the spatial coordinates change with time since we are focusing on a fixed particle which moves from one point to another. In addition to this, there is the explicit time dependency. The differential change in temperature of the particle can be written as

$$dT = \frac{\partial T}{\partial x}dx + \frac{\partial T}{\partial y}dy + \frac{\partial T}{\partial z}dz + \frac{\partial T}{\partial t}dt \tag{4.13a}$$

or

$$\frac{dT}{dt} = \frac{\partial T}{\partial x}\frac{dx}{dt} + \frac{\partial T}{\partial y}\frac{dy}{dt} + \frac{\partial T}{\partial z}\frac{dz}{dt} + \frac{\partial T}{\partial t} \tag{4.13b}$$

86 Introduction to Chemical Engineering

$$= v \cdot \nabla T + \frac{\partial T}{\partial t} \qquad (4.13c)$$

This relates to the rate of change of temperature T measured using the two approaches. The left-hand side gives the rate of change of temperature of a particle (or a fixed infinitesimal mass) which moves along with the flow. The last term on the right-hand side, the partial derivative with respect to time gives the rate of change at a fixed position in space which is occupied by different particles at any instant. This is the rate of change of temperature of a fixed infinitesimal volume. The partial derivative is also called the *local derivative* or the *Eulerian derivative* while the total derivative [on the left of Eq. 4.13(c)] is also called the *Lagrangian derivative* or the *material derivative*. These correspond to the rate of change measured using the two different approaches in analysing a situation.

The above equation when applied to the velocity of a particle results in

$$\frac{Dv}{Dt} = v \cdot \nabla v + \frac{\partial v}{\partial t} \qquad (4.14)$$

This is called the *Euler's acceleration formula*.

The difference in the two derivatives is best seen in the following example. Consider the flow of a field in a converging pipe (Figure 4.3). Let us assume the velocity at any axial position is spatially uniform in the transverse direction. Since the rate of inflow of liquid at section 1 where the area is A_1 must equal the rate of outflow of liquid at section 2 where the area is A_2 it follows from the conservation of mass that

$$v_1 A_1 = v_2 A_2$$

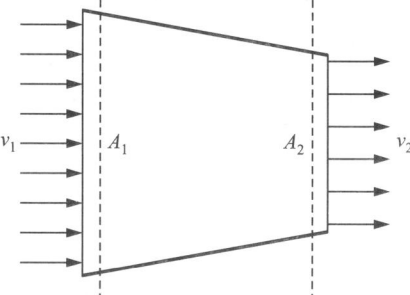

Figure 4.3 One-dimensional flow through a converging pipe, showing particle acceleration under steady conditions.

This follows from the fact that the rate at which mass comes in must equal the rate at which mass leaves a control volume. The control volume is indicated by a dashed line. Here we have used the fact that liquids are incompressible, i.e. their density is a constant. At each point the velocity is a constant since the flow is steady $\partial v/\partial t = 0$. The particles, however, experience acceleration as the velocity increases from one point to another, in going from point 1 to 2 in the pipe. This causes $dv/dt \neq 0$. From Euler's acceleration formula $dv/dt = v(\partial v/\partial x)$.

Here we have a situation where the material derivative is not zero since the particles accelerate but the partial derivative is zero as the flow is steady.

The extension of this relationship which is valid for a point to a finite region in space is called the *Reynolds Transport Theorem*. This will not be formally derived here. We will give an insight into how this theorem arises for macroscopic open systems or a finite-sized control volume.

Macroscopic Control Volume

Unsteady State: Ordinary Differential Equations

Let us start with an example. Consider a bucket in which water is being filled. Here we apply the conservation of mass for the bucket. In view of the mass of water flowing in there is accumulation of mass of water in the bucket [Figure 4.4(a)]. So if we look at the bucket of water it would appear that mass is being created. In reality we are not creating mass. So where does the paradox arise. We are focusing our attention on a fixed region in space (the bucket) and mass is entering through the control surface. Thus, if q is the volumetric flow rate of water, then the mass flow rate of water is ρq or m. This must be equal to the rate at which mass of water is accumulating inside the bucket. Thus, for the bucket, we have

Rate of accumulation of water = Rate at which water is entering   (4.15)

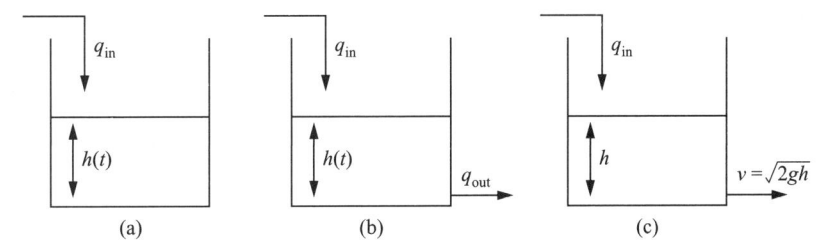

Figure 4.4 The dependence of water level in a tank: (a) when only inflow is present, (b) inflow and a constant outflow is present, and (c) with inflow and a height dependent outflow velocity.

We can generalize this to a system in which we have an inflow (due to a pipe carrying water in) as well as an outflow (due to a pipe from which water flows out) [Figure 4.4(b)]. The above equation can be generalized to

Rate of accumulation of water
 = Rate at which water enters in − rate at which water exits

Hence, we can say that the accumulation in a control volume is equal to the net inflow across the control surface. This example represents a dynamic system where a dependent variable changes with time. Here the height of the liquid inside the tank varies with time. For a tank with cross-section A and when the height of the fluid is h, the above equation can be written as

$$A\frac{dh}{dt} = q_{in} - q_{out} \tag{4.16}$$

Here q represents the volumetric flow rate.

The flow rate coming out of the tank, usually is not a constant. It is dependent on the height of the water in the tank. You will see later that the velocity and, hence, the flow rate of the stream leaving the bottom of a tank has a square root dependency on the height of water in the tank [Figure 4.4(c)]. If this is explicitly included in the equation, we have

$$A\left(\frac{dh}{dt}\right) = q_{in} - \alpha\sqrt{h} \tag{4.17}$$

Here α represents a proportionality constant. This is a nonlinear ordinary differential equation which needs to be solved. It is nonlinear since the dependent variable h occurs to a power other than unity (0.5) on the right. The solution to this is subject to an initial condition at $t = 0$, $h = h_0$. Physically, the initial condition specifies the height in the tank at the initial instant or beginning of the experiment or operation. Mathematically, it is a condition which is necessary to specify a unique solution. Without the initial condition the solution is determined within an arbitrary constant. The solution of Eq. (4.17) gives how h changes with time, i.e. how h evolves with time from the initial value to the steady value.

Equation (4.16) is an ordinary differential equation which describes how the height of water in the tank h changes with time for a given q_{in}, q_{out}. When $q_{in} > q_{out}$ the height in the tank increases linearly with time. Can you sketch the curve? Is the solution unique? Does this solution hold for all time t?

When $q_{in} < q_{out}$ the height decreases linearly with time and if you integrate long enough the height would be negative. Since the dependent variable h has a physical meaning it does not make sense to have a negative height [Figure 4.5(a), (b)]. Hence the integration has to stop when h becomes zero since after that time it remains at zero. A mathematician sees only the equation that is written and would blindly integrate it to get a solution even continuing when $h < 0$. As an engineer one needs to understand that the equation has a physical relevance and use that perspective in the interpretation and analysis of the results.

When the tank is at a steady state the height of the liquid does not change with time and, consequently, there is no accumulation of mass in the system.

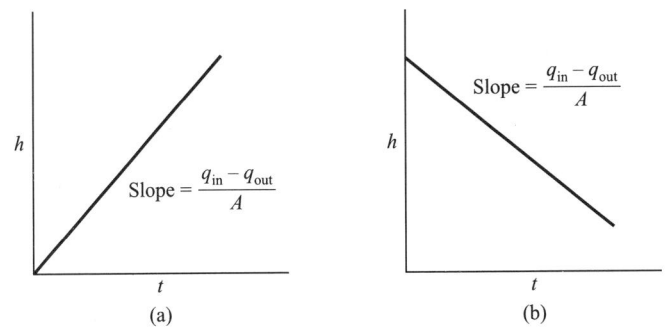

Figure 4.5 Variation of height with respect to time in a tank: (a) $q_{in} > q_{out}$ and (b) $q_{in} < q_{out}$.

At steady state the accumulation term equals zero and the conservation law reduces to

Rate at which water enters in − Rate at which water leaves out = 0

This equation can be solved for the case when the outlet flow rate depends on h as in Eq. (4.17) to determine the height of the water level in the tank at steady state. This yields $h = (q_{in}/\alpha)^2$. The solution of the differential equation describes how the height in the tank varies from the initial value to the final steady state. Can you draw the profiles of h with time as predicted by Eq. (4.17). What is the difference between these curves and the linear curves obtained when q_{out} is a constant?

A typical picture of this evolution is shown in Figure 4.6. Two cases are depicted corresponding to the initial height being above and below the steady state. For the former (latter) the height decreases (increases) monotonically with time till it reaches the steady state. The solution to nonlinear equations is obtained numerically through computer programs. The code in MATLAB which integrates the differential Eq. (4.17) is given in MATPROG 2 in the Appendix.

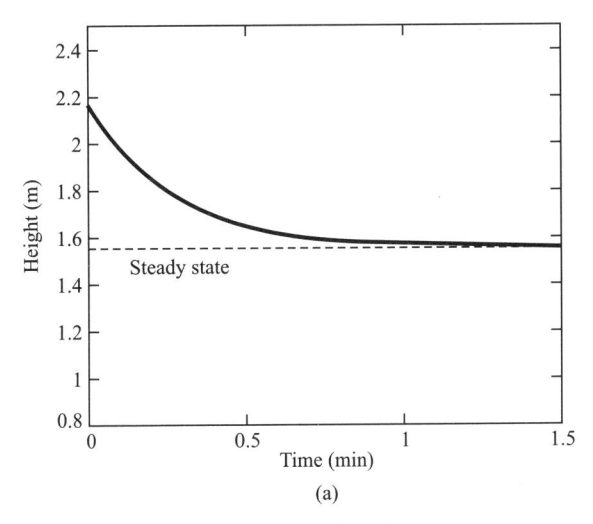

Figure 4.6 Contd.

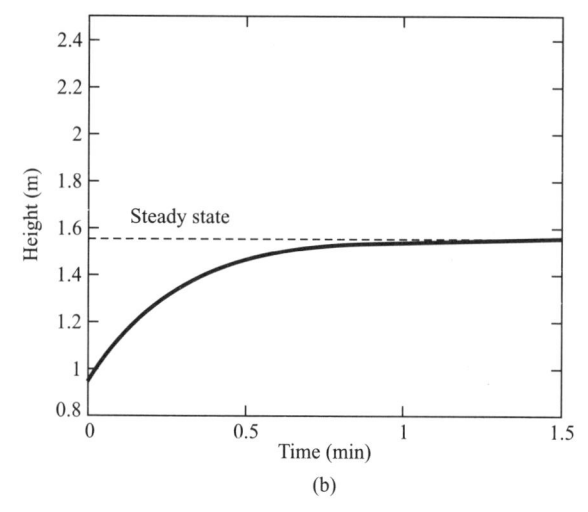

[Graph showing height vs time curve approaching steady state from below, with y-axis "Height (m)" from 0.8 to 2.4, x-axis "Time (min)" from 0 to 1.5, dashed line labeled "Steady state" at about 1.6]

(b)

Figure 4.6 Variation of h with time t as depicted by Eq. (4.17). (a) $h_0 > h^{ss}$, (b) $h_0 < h^{ss}$ where h_0 is the initial height. The parameters used in the simulation can be found in the program in the Appendix.

Stability Issues in a System

In the tank problem just discussed it was seen that the outlet flow rate would be proportional to the square root of the height in the tank. This implies that we can find the steady state by setting the time derivative in the mass balance to zero. In typical operations of plants there are frequent disturbances. For instance, the inlet flow rate to a tank may not be a constant but may vary around it. This can cause changes in the height of water in the tank. When disturbances are present the height deviates from the steady state. One of the questions a chemical engineer would like to address is how does the system respond to these disturbances. Disturbances result in a change in the dependent variable, in this case the height of water in the tank.

Consider the tank at steady state, where the inflow equals the height dependent outflow. A disturbance in this system could be a temporary (a very short duration) blockage of a pipe, either in the inlet or the outlet in the above example. This can result in a momentary increase or decrease in the flow rate before going back to the steady value. This will cause the liquid height to change from the earlier value. Alternatively, suppose a mug of water is removed from the tank at steady state. This can also be a disturbance. As a result, the height of water in the tank deviates from the steady state. The question asked is how does the height in the tank change (evolve with time) if the system is left to itself, i.e. after the source of the disturbance has been removed.

Let us analyse this physically now. Taking a mug of water out results in lowering the height in the tank. The system is not at steady state since the inflow rate and outflow rate are not balanced anymore and the height in the tank evolves with time. The question is: will the system reach the original steady state value (from which it was disturbed) again if left to itself? Let us see if we can physically reason and determine what is likely to happen in this simple situation. Removal of water results in the height in the tank being lower than the steady-state height. As a result, the outlet flow will be lower and the inlet flow which is independent of the tank height is a constant. Now there is a net influx to the tank and the height increases. If you were to add a mug of water the height in the tank would increase and deviate from the steady state value. Now the reverse happens and there is a net outflow of water from the tank (as the outflow is more than the inflow) and, hence, the system again evolves back to the height at steady state. The steady states of the system which are such that when disturbed and left to themselves evolve back to the original steady state are called *stable steady states*. The stability of the steady state here is an intrinsic characteristic of the system and hence we do not need to control it.

There are other systems which do not have a stable steady state. An example of this is a reactor sustaining an exothermic reaction. Here the steady state is characterized by the concentration and temperature in the reactor. Consider a disturbance which results in an increase in the reactor temperature. This results in a higher rate of reaction. This in turn will generate more heat by exothermicity and the temperature will keep on increasing and will not return to the original steady-state value. Here as a result of the disturbance the temperature deviates further away from the steady state. This is a very simplistic argument used to illustrate instability. The above system has steady states which are stable as well. In reality the stability is governed by a combined influence of both the concentration and temperature. A mathematical framework is necessary for determining the stability of the steady states of these systems characterized by the interaction of different variables.

The area of process control discusses the methodology to determine if a steady state is stable or unstable for systems where the interactions are more complex as in the example of the exothermic reaction. A rigorous generalized mathematical framework is established for determining the stability. Most processes are complex in the sense that the different variables interact with each other. For instance, the height, concentration, and temperature may interact with each other in a tank and may have opposite influences and it would be difficult to determine if the system is stable or not on the basis of simple physical arguments like what we have made above. In a process plant with recycle streams this interaction becomes even more involved. Hence there is a need of a mathematical theory to be developed to determine the stability of a state and determine the methods of controlling unstable steady states. Chemical engineers are involved with the design of these control systems.

Conservation of Mass: Applications

Let us come back to the applications of the conservation of mass equation. Consider a fluid flowing through a sudden contraction. Here the cross-sectional area of the pipe changes sharply at a point as shown in Figure 4.7.

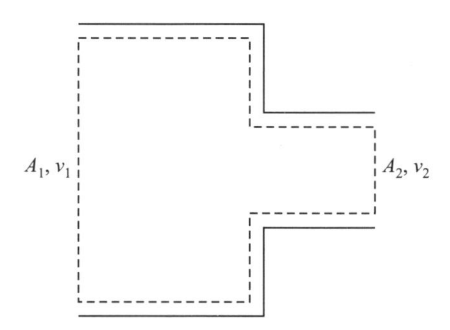

Figure 4.7 Flow through a sudden contraction.

The mass-flow rate in one pipe is known. Our objective is to determine the mass-flow rate through the other pipe and the velocity of the fluid through it. For simplicity we assume the velocity is uniform across the pipe. We choose a fixed control volume as shown by the dotted line and this then yields that the accumulation is zero at steady state. We obtain

Rate of influx across the control surface
= Rate of efflux across the control surface

This gives $\rho_1 A_1 v_1 = \rho_2 A_2 v_2$.

This equation states that the mass-flow rate through each of the pipes is the same, and this can be used to determine the velocity through one pipe given the velocity through the other.

As an extension of the above problem, consider a pipe network shown in Figure 4.8. Here the pipe bifurcates into two pipes. Now we have

$$m_{in} = m_{out1} + m_{out2} \tag{4.18}$$

In terms of the density (ρ), area (A), velocity (v), this gives

$$\rho_{in} A_{in} v_{in} = \rho_{out1} A_1 v_1 + \rho_{out2} A_2 v_2 \tag{4.19a}$$

For incompressible liquids, all the densities are equal and this simplifies to

$$A_{in} v_{in} = A_1 v_1 + A_2 v_2 \tag{4.19b}$$

Is this similar to something you have seen before? In fact, this is analogous to *Kirchhoff's first law* (in the context of resistor networks) for current which states that the sum of the currents entering the node must equal the sum of the currents leaving the node. The current here is equivalent to the mass-flow rate while the current density is equivalent to the velocity. The above equation can then be used to obtain velocity inside each pipe, provided we know the other variables. The conservation of mass equation is also called the *equation of continuity*.

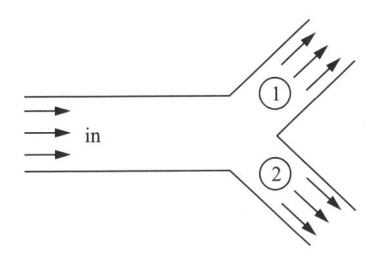

Figure 4.8 Flow through a bifurcation.

The *law of conservation* we have discussed so far is for total mass or material mass. However, in chemical engineering we are sometimes interested in applying the law for conservation of mass to a particular species. This is particularly important when we deal with chemical reactions. Here the interest is in understanding how much of a species has been produced or consumed by the reaction.

Conservation of Mass Reacting Systems

Consider, for instance, a batch reactor in which a liquid phase reaction is occurring. For simplicity we assume the reaction to be an isomerisation reaction which can be written as $A \to B$, where A is the reactant and B is the product. Here as the reaction progresses the total amount of mass is constant in the reactor (assuming the reaction is conducted in a closed system, i.e. no products can leave the reactor). However, if we look at the mass of a species, say, the reactant A or a product B it changes with time in the batch reactor. We now have a situation wherein the mass of species changes in a fixed total material mass or a fixed collection of molecules. The rate at which this changes is given by the reaction rate. Let r be the reaction rate defined as the rate at which a particular species is generated per unit volume. Its unit is kg mol/s-m^3 (chemical engineers usually work with moles). The rate of generation of a species (kg mol/s) in the reactor is given by rV, where V is the volume of the liquid in the reactor. Here we are assuming that the rate of reaction is the same at all points in the reactor.

In the batch reactor, if we were to focus on the total mass of the reactor, then the rate of accumulation of mass in the reactor would be zero. When we follow the change in the concentration of a species we would see that the rate at which the species accumulates in the reactor is equal to the rate of generation of the species. There is no inflow or outflow since we have a batch system. Consequently, in a batch reactor we have

The rate of accumulation of species B
= Rate of generation of species B (4.20)

This yields

$$V\frac{dC}{dt} = rV \qquad (4.21)$$

where r represents the reaction rate which is a function of concentration.

The left-hand side of Eq. (4.21) represents the accumulation term. And the right-hand side represents the rate of generation term for a particular species which is the rate of generation for a constant mass system. This term represents the rate at which a particular species is generated or consumed (accordingly the sign will be positive or negative) by a reaction even when there is no flow across the control surface. When there is, in addition to the reaction, an inflow and an outflow from the reactor as in a continuous reactor we extend the law of conservation of mass to

Rate of accumulation = Inflow − Outflow + Rate of generation

Here the accumulation term represents the rate of accumulation in a control volume. The inflow (*outflow*) term represents the rate at which mass enters (leaves) the control volume across the control surface and the rate of generation term is the rate at which mass is generated in a control mass which occupies the control volume at the same instant. When we apply the conservation of mass to the total mass the generation term is zero and when we apply it to a species this is equal to the reaction rate.

The above law is the extension of the conservation law to open systems. This can be formally derived for various quantities like mass, momentum, and energy. These laws are used in different applications in chemical engineering. When applying these to a specific system, the different terms take on different expressions. A chemical engineer learns how each term is represented in various contexts.

The above relationship can be similarly extended to conservation of momentum. This yields

Rate of accumulation of momentum in a CV
= Rate at which momentum enters in
− Rate at which momentum exits
+ Rate at which momentum is generated in the control mass
occupying the control volume

Here the generation term is associated with the rate of generation of momentum for the mass which occupies the control volume. This from *Newton's second law of motion* can be represented by the net forces acting on the control mass present in the control volume.

The above equation is the generalized conservation law which you will come across over and over again in different contexts. This can be applied to mass, momentum or energy. A formal derivation of this which relates changes in a control volume (accumulation) to control mass (generation) occupying the control volume is called the *Reynolds Transport Theorem* which you will see later on in the curriculum.

Application to a Flash Unit: Nonlinear Algebraic Equations

A flash system is now considered to illustrate the principle of conservation of mass. The flash unit is a simple possible unit which can be used for continuously separating components in a mixture on the basis of the difference in boiling points. A schematic representation of the process is shown in Figure 4.9. Here a stream of liquid containing two components A, B is flowing through a pipe. The total molar flow rate is F mol/s and its composition, i.e. molefraction is z_a, z_b. The stream is heated which causes a rise in temperature. This hot stream then passes through a valve or a nozzle across which there is a significant drop in pressure. The lowering of pressure causes the stream to partially vaporize. The vapour stream has a flow rate of V mol/s and a composition of y_a and y_b. Similarly, the liquid stream has a flow rate of L mol/s and a composition of x_a and x_b.

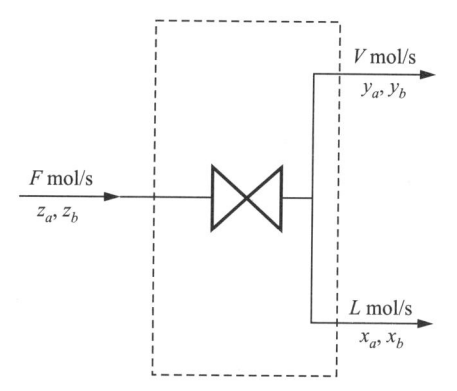

Figure 4.9 Schematic diagram of a flash through throttling valve.

Our objective is to determine the compositions of the two streams leaving the flash unit, and their flow rates. We have a nonreacting mixture and since the process is continuous there can be a steady state. We consider the system to be at steady state. Consequently, the accumulation term is zero here. It is assumed that there is no reaction occurring and, consequently, there can be no generation of any species and the generation term is zero. Hence the conservation of mass reduces to

Rate of species coming into the flash = Rate of species leaving the flash

The overall mass balance equation for the system is

$$F = V + L \quad (4.22)$$

The above equation states that since there is no reaction the total number of moles entering must be equal to the total number of moles leaving at steady state. We are now working in moles as chemical engineers should.

For species A, we similarly have

$$Fz_a = Vy_a + Lx_a \qquad (4.23a)$$

What are the variables which have to be determined? There are a total of four variables V, L, x_a, y_a. The other two mole fractions are dependent on these since for a binary system the mole fractions in each phase must sum to unity.

Hence, we have two equations and four unknowns. So to determine the performance of the system, we need two more equations.

It appears that one more equation can come from a species B balance which is

$$Fz_b = Vy_b + Lx_b \qquad (4.23b)$$

However, this equation is not an independent equation and can be obtained by subtracting the total or overall mass balance from the species A balance. Consequently, no new information is present in this and we cannot use it. We have only two equations arising from the mass balance. The number of independent equations which can be obtained from a mass balance is equal to the number of components in the system.

To determine or predict the system performance we have to determine two more equations or we need to specify two of the four variables and solve for the remaining two.

To determine the system behaviour the number of unknowns must equal the number of equations. If the number of equations is more, then we will not be able to satisfy all of them and may violate some equations. We have what is called an *over-determined system* (as seen in the source apportionment problem in air pollution earlier in this chapter).

On the other hand, if the number of equations is less, then we cannot determine all the variables uniquely. We have now an under-determined system. The number of variables which we have to specify to determine the system behaviour uniquely is called the *degrees of freedom of the system*. The degrees of freedom of a system is given by the difference in the number of unknowns and the number of equations. So in the above case it appears like the degrees of freedom is two, i.e. we need to specify two variables out of the four remaining to find a solution. Alternatively, we have to specify two more equations based on physical principles.

For the flash unit two more equations can be obtained by assuming the exiting vapour and liquid streams to be in equilibrium. This assumption generates a relationship between the compositions of the two exiting streams. While several complex expressions are available, which you will see later in your course on Chemical Engineering Thermodynamics for simplicity it is assumed that the equilibrium follows Raoult's law. Now the equilibrium relation is of the form

$$y_a = K_a x_a, \quad K_a = \frac{P_a^{sat}}{P} \tag{4.24}$$

Here P_a^{sat} is the saturation pressure of A and is a function of temperature and P represents the system pressure. Writing this for both species (A, B) and considering the mass balance for the overall system and the individual species we have four equations in four unknowns which can be used to determine the system behaviour. While the validity of Raoult's law can be questioned, this has been used only to illustrate specifically how the assumption of equilibrium between the two streams can be used to solve the problem. There are other relationships which can be used to determine the equilibrium compositions depending on the actual system being analysed.

Here a tacit assumption made is that the temperature in the flash unit is known. This allows us to evaluate the saturation pressure and hence K_a and K_b. However, the temperature of the flash unit may be unknown and the flash may be operated under a fixed heat duty. Here the temperature is determined from the energy balance. This can be written as

Rate of heat entering = Rate of heat leaving the control volume

Here the steady-state operation is assumed and the generation term is absent since there is no reaction. This results in

$$Q + Fh_f = Vh_v + Lh_l \tag{4.25}$$

Here h represents the specific enthalpy and the subscripts f, v, l are used to identify the feed, vapour, and liquid streams. This energy balance can be used to determine T for a given heat load Q or to determine Q for a fixed and desired flash temperature.

The set of equations we have written down are nonlinear algebraic equations. The *Antoine equation* or the *Clausius–Clapeyron equation* is normally used to calculate the saturation pressure. It has an exponential dependency on temperature $P^{sat} = \exp(A - B/T)$. The resulting system of nonlinear algebraic equations cannot be solved analytically. They are nonlinear as they have the product of two dependent variables (molar flow rate and composition). They also contain an exponential term as saturation pressure depends exponentially on temperature. How do we solve these equations? They need to be solved numerically which means computer programs have to be written. The nonlinear algebraic equations have to be solved iteratively and the techniques for solving these equations will be taught in the course on *computational techniques* or *numerical methods*. The techniques are based on mathematical concepts and computer programs are a way to implement these concepts.

It is seen that the application of conservation laws yields different kinds of mathematical equations: linear algebraic equations, nonlinear algebraic equations or linear and nonlinear differential equations. Methods of solving these equations which vary in the levels of complexity are hence an integral

98 *Introduction to Chemical Engineering*

part of chemical engineering. Not only are we interested in the methods of solving the system of equations but as engineers it is important to ensure that the solutions are physically realistic, since the equations have a physical meaning.

Physically Admissible Solutions

The meaning of a physically realistic solution is now explained with an example. One of the classical reactors used in the chemical process industry is a continuously stirred tank reactor (CSTR). Here the reactants are fed continuously to the reactor. The contents of the reactor are well stirred and the partially converted reactants (a mixture of reactants and products) leave the reactor as shown in Figure 4.10.

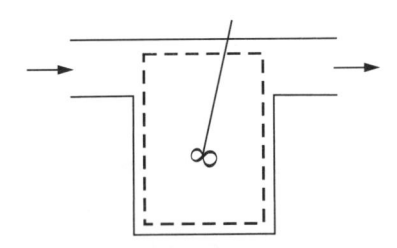

Figure 4.10 Schematic diagram of a CSTR.

The following assumptions are made for the analysis of the CSTR:

1. The contents of the reactor are well stirred, so composition and temperature inside are spatially uniform, i.e. they are the same everywhere.
2. Conditions of concentration and temperature in the exit stream are identical to those prevailing in the reactor.

To analyse this system and to predict the performance of the reactor, the conservation of mass is used. The control volume is chosen as the region enclosed by a dashed line. The dashed line is the control surface. The volume of the reactor, i.e. the height of the reactor contents is assumed to be a constant. We now carry out a mass balance for species (A) in the reactor. A second-order reaction of the form $A \rightarrow B$ is assumed to take place in the reactor. This could be an *isomerisation reaction*.

The mass balance equation applied to this open system is

Rate of accumulation of species A
= Rate at which species A comes in
− Rate at which species A leaves the control surface
+ Rate of generation of species A due to reaction

The volumetric flow rate entering and leaving the system is assumed to be equal to q m³/s. The concentration in the feed and exit stream is given by

C_{Af} and C_A. Here the inflow term represents the rate at which moles of A enter the control volume = qC_{Af} mol/s.

The outflow term represents the rate at which moles of A leave the control volume = qC_A mol/s.

Since the volumetric flow rates entering and leaving the reactor are equal, the volume of the reactor does not change with time (neglecting the density changes). The accumulation term represents the rate of change in the number of moles of A in the reactor. Since the reactor volume is a constant and the concentration inside the reactor is the same as that in the exit, we have

$$\frac{dN_A}{dt} = V\frac{dC_A}{dt} \qquad (4.26)$$

The generation term represents the rate at which species (A) is generated in the material mass which occupies the control volume at an instant.

As the reaction is the second-order, this is given by $-VkC_A^2$.

The negative sign arises as species A is consumed by the reaction and is not generated by it. Substituting these expressions, we obtain

$$V\frac{dC_A}{dt} = q(C_{Af} - C_A) - VkC_A^2 \qquad (4.27)$$

For a given reactor volume (V), flow rate q and reaction kinetics (k), the solution to this equation represents how C_A evolves with time.

As time increases and tends to infinity the concentration of A, C_A approaches a constant value and this is the steady state. The steady state can also be found by setting the left-hand side of the expression to zero, this gives us

$$0 = q(C_{Af} - C_A) - VkC_A \qquad (4.28)$$

This equation being a quadratic has two roots:

$$C_A = \frac{-1 \pm \sqrt{1^2 - 4\frac{VkC_{Af}}{q}}}{2\frac{Vk}{q}}$$

It is clear that although the quadratic equation admits two solutions; one of them is always negative and hence is physically inadmissible or unrealistic. We hence have only one solution corresponding to the '+' sign in front of the discriminant.

The exercise of writing down conservation laws to determine the system behaviour is called *modelling*, and the process/exercise of solving the equations thus arising is called *simulation*. This approach allows us to predict the system performance for different reactor sizes, flow rates and reactor kinetics without having to carry out experiments. While this approach is sufficiently general and has a strong scientific basis, the accuracy of the predictions is determined by how well we understand the system. The accuracy of the predicted steady-state concentration of the CSTR above is determined by whether the assumptions are valid or not. Some of these assumptions include: spatial uniformity inside the tank, exit concentration equal to reactor concentration, second-order kinetics, etc. Should these assumptions be invalid, our predictions can be very inaccurate.

Another way of analysing a system is by carrying out experiments. Here we need to carry out experiments on a set-up and obtain insight into the performance. Here we would get realistic information, however it would be difficult to get data for other operating conditions like a different volume of reactor or a different flow rate through the reactor. Using this approach to find the performance for a different set of conditions we would have to redesign and redo the experiments. The results obtained using this experimental method though accurate cannot be generalized to other conditions easily.

Ideally, it is desirable to use a two-pronged approach involving experiments, on the one hand, and modelling and simulation, on the other hand. Each would help independently validate the results of the other approach. One goal is to find a reliable model which can be applied to predict system behaviour under new operating conditions. The model would be valid and gives accurate results when the operating conditions are in the range of parameters where the experiments were conducted.

The systems we have seen so far are called *lumped systems*. Here the spatial variations in the concentrations of various species, are neglected. The system is assumed to be spatially uniform, i.e. the concentration is uniform across the tank. We can hence view the equations written above as laws applied to macroscopic system.

An Example Showing a Partial Differential Equation

There are several situations in chemical engineering where we have to solve partial differential equations. Here the dependent variables evolve with space and time or in more than one spatial direction. Such systems are called *distributed systems*. In analysing these systems again we use the same principles of conservation but they are applied to infinitesimally small or microscopic control volumes. This is in contrast to the case so far where the principles have been applied to control volumes which are macroscopic and where the variables, intensive properties were assumed to be spatially uniform inside.

We take the example of conservation of energy to illustrate this. Consider a solid rod of copper which is at a initial uniform temperature of 80°C. The curved surface as well as one end is assumed to be insulated so that it does not lose heat to the surroundings across these surfaces. The other end of the rod is suddenly dipped in an infinite expanse of cold water at a temperature of 30°C. The infinite expanse ensures that the temperature of the water does not increase beyond 30°C even after it gains heat from the rod (Figure 4.11).

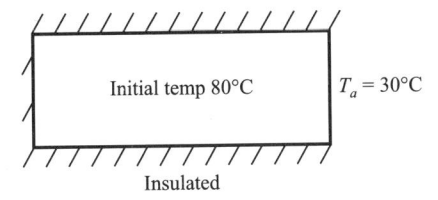

Role and Importance of Basic Sciences in Engineering **101**

Figure 4.11 Hot rod losing heat to the ambience through one end.

If we assume that the temperature depends only on axial coordinate, can you describe through a sketch how the temperature will change with time? What will be the final steady temperature if we were to wait for a sufficiently long time?

Figure 4.12 depicts schematically how temperature inside the rod varies with axial position at different instants. At $t = 0$, the temperature is uniform at 80°C. As time increases, the temperature decreases. The decrease is rapid near $x = L$. The end at $x = 0$ decreases slowly. If we wait for a sufficiently long time, the temperature everywhere in the rod will be uniform at 30°C.

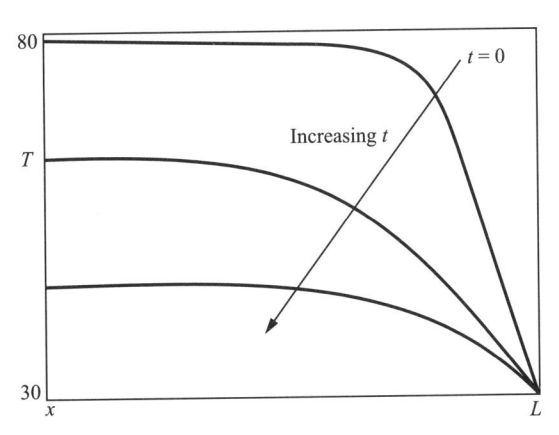

Figure 4.12 Temperature inside a rod which is at 30°C at $x = L$ and insulated at $x = 0$ as time increases the temperature at each point decreases.

As an engineer, we would like to obtain quantitative estimates of the profiles. How can we best describe this? For this we need to write down the energy balance equation and solve it. Here the interest is in the spatio-temporal variation of temperature. To determine this, an energy balance is taken over a thin shell of thickness dz this extends from z to $z + dz$ as shown in Figure 4.13. For the sake of generality, heat loss from the curved surface is also considered now.

The control volume is chosen as the thin shell. The rate of heat influx by conduction across the left face is $-kA(\partial T/\partial z)_z$. This is the rate at which heat enters the infinitesimal control volume and arises from the *Fourier law of heat conduction*.

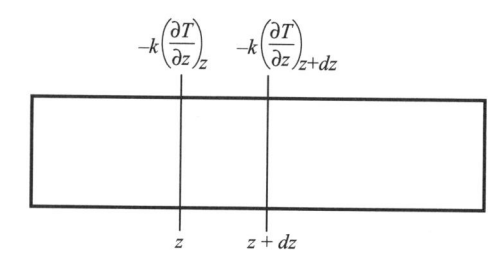

Figure 4.13 Energy balance across a shell of thickness dz.

The rate of heat efflux by conduction across the right face is $-kA(\partial T/\partial z)_{z+dz}$. This is the rate of heat loss from the right surface (which is present at $z + dz$). Heat is also lost through the circumferential area to the ambient. This heat loss is proportional to the temperature difference and the surface area. Here h is the proportionality constant and is called the *heat transfer coefficient*. This constant depends on the conditions prevailing in the ambient, i.e. whether the ambient fluid is flowing and the properties of the ambient fluid. For example, if the fluid is stationary h will be low and but if the fluid flows at a high velocity around the surface h will be high.

This is also a heat efflux term and is given by $h2\pi R dz(T - T_a)$.

The rate of accumulation of heat is $\rho c_p A dz(\partial T/\partial t)$.

The energy balance equation reduces to

$$\rho c_p A dz \frac{\partial T}{\partial t} = -kA\left(\frac{\partial T}{\partial z}\right)_z - kA\left(\frac{\partial T}{\partial z}\right)_{z+dz} - h2\pi R dz(T - T_a)$$

Dividing by dz and taking the limit as $dz \to 0$ yields

$$\rho c_p \frac{\partial T}{\partial t} = \lim_{\Delta z \to 0} \frac{k\left(\frac{\partial T}{\partial z}\right)_{z+dz} - k\left(\frac{\partial T}{\partial z}\right)_z}{\Delta z} - \frac{h2\pi R(T - T_a)}{\pi R^2}$$

$$\rho c_p \frac{\partial T}{\partial t} = k\left(\frac{\partial^2 T}{\partial z^2}\right) - \frac{2h}{R}(T - T_a) \qquad (4.29)$$

This is a partial differential equation whose solution describes how the temperature in the rod will change as a function of z and time t. If the curved surface is insulated, then no heat loss can take place across it. This can be achieved by setting h to zero in Eq. (4.29).

In Eq. (4.29), the radial distribution of temperature is neglected. This is valid when the radius of the cylinder is very small so that heat conduction smears out any variations in the radial direction. For small diameters the heat transfer by convection from the surface to the ambient is much slower than the heat transfer by conduction from the centre to the surface of the rod.

To find a unique solution to this equation initial and boundary conditions have to be specified. Since it is first order in time and second order in space, one condition in time and two in space are required. The condition in time specifies the condition at time $t = 0$ and is called the *initial condition* and the conditions in z specify the conditions at the two ends of the rod and are the boundary conditions. These come from the physical problem at hand and the conditions prevailing.

Thus, the initial condition is at $t = 0$, we have $T = 80$ for all $z > 0$, since the rod is at a uniform temperature of 80°C at the beginning.

The boundary conditions are:

At $z = 0$, we have $\partial T/\partial z = 0$ for all t, if this end is insulated and heat cannot be lost. This condition says that the slope of T will be zero at $x = 0$. This has to be satisfied at all instants. The temperature profiles in Figure 4.12 attest to this.

At $z = L$, we have $T = 30$°C for all t, if this end is dipped in water at 30°C.

The solution to this partial differential equation describes how the temperature in the rod varies with space (z) and time. This generates the spatial dependency at different instants as shown in Figure 4.12. It can be seen that these profiles satisfy the two boundary conditions, i.e. at the right end the temperature is 30°C and at the left end the temperature derivative is zero.

Optimization Problems in Chemical Engineering

Consider a vessel in which we have a heat source, where heat is generated at the rate of Q W/m^3. Such a system can be realized when you warm water with an electrical heating coil. If the vessel is at a steady state, then the rate of heat generation must equal the rate at which heat is being lost to the ambience. The rate of heat loss to the ambience is determined by (in fact is proportional to) the temperature difference between the vessel contents and the ambient temperature (T_a) and the surface area available for the heat transfer. The temperature in the vessel is assumed to be spatially uniform. The proportionality constant is called the *heat transfer coefficient h* as before. This gives us

$$QV = hA(T - T_a)$$

For a given volume of the vessel V and a fixed heating rate Q, it can be seen that the vessel with a larger surface area will have a lower steady-state temperature T. While this is quantitatively intuitive the exact temperature can be determined from the above energy balance.

For a fixed volume of the vessel the surface area available for heat transfer can be changed by altering the geometry or vessel shape. The sphere has the lowest surface area for a volume and hence temperature of the liquid in a spherical vessel would be highest. However, keeping a spherical vessel in place could be a challenge (it could just roll off) and it may be more practical to have a cylindrical vessel. A cylinder of the same volume would have a higher surface area across which heat transfer occurs and this can result in a lower temperature.

We know that $V_s = (4/3)\pi R^3$, $S_s = 4\pi R^2$ represent the volume and surface area of sphere of radius R. A cylinder has volume of $\pi r^2 h$ and a total surface area of $(2\pi rh + 2\pi r^2)$. The radius r and height h of the cylinder can be chosen to have the same volume as that of the sphere. This ensures that the volume of the liquid inside the two geometries is equal for a fair comparison.

Imposing this constraint, we obtain

$$\pi r^2 h = \frac{4}{3}\pi R^3$$

Using this condition, the surface area of the cylinder is written in terms of r, the radius. We eliminate h, the height using the above constraint to obtain the surface area of the cylinder as $2\pi(4/3)(R^3/r) + 2\pi r^2$. The surface area is a function of only one variable now r and its optimum can be found by setting the first derivative to zero. This yields

$$-\frac{4R^3}{3r^2} + 2r = 0$$

or

$$r = \left(\frac{2}{3}\right)^{1/3} R$$

The second derivative is $\frac{8R^3}{3r^3} + 2 > 0$ and hence the optimum found is a minimum. The corresponding h value is $2\left(\frac{2}{3}\right)^{1/3} R$ and the surface area of the cylinder is

$$= 2\pi \left(2\left(\frac{2}{3}\right)^{2/3} R^2 + \left(\frac{2}{3}\right)^{2/3} R^2 \right)$$

$$= 6\pi \left(\frac{2}{3}\right)^{2/3} R^2$$

Here we have found the minimum surface area of a cylinder for a fixed volume. For the choice of the radius and corresponding height the temperature inside the cylindrical vessel will be the highest. This minimum surface area is still higher than that of a sphere for the same volume. This is a practical example which has been solved by changing the geometry of the system (the length and diameter of the cylinder) and which gives rise to a constrained optimization problem. The constraint arises from the condition that the volume of the cylinder and the sphere have to be equal. Similar optimization problems arise in different contexts in chemical engineering and their solutions require more detailed mathematical techniques.

Exercises

1. In the source apportionment problem related to determining contributions of sources of pollution, a student suggests resolving the problem by measuring $n - 1$ elements and the total mass when there are n sources present. This way the number of equations and unknowns match and the system of equations will have a unique solution. Do you think this is a good approach or should there be measurements of more elements?

2. In the source apportionment problem you determine the solutions and you find that some of them are negative. (Some of the contributions are negative).
   (i) What is your comment? Is this permissible?
   (ii) In your opinion, what causes this? How can you fix this? Can you suggest some ways?
   (iii) Is it possible to determine only those solutions which are positive? Can we impose this as a constraint?

3. You have a quantity of sodium carbonate with you on a filter paper. Your friend Ram adds some amount of sodium bicarbonate to this mistaking the salt to be the bicarbonate. Another friend Shyam adds some amount of sodium carbonate to this. Can you find out the individual amount (mass in grams) of the salts added by each person from the composite sample? If you can, what information would you need. Explain.

4. A mixture of calcium sulphate, potassium sulphide and potassium oxide is given to you. You are asked to find out the amount of each of the compounds present in the mixture. The mixture is analysed quantitatively for the amount of each elements. The amount of calcium is found to be 20%, potassium is 30%, oxygen is 10% and the remaining is sulphur. Find the mass of each compound present in the mixture.

5. Consider the equation $-x + \delta e^x = 0$. This represents the energy balance of a zeroth order reaction in a CSTR. Find the solutions to this for $\delta = 1$, $\delta = 5$. Draw the graph of $y = x$ and $y = \delta e^x$ and examine its point of intersection.

6. Rework the constrained optimization problem of the cylinder when heat loss is possible through the curved surface and only one of the ends.

7. Which reactions are inherently unsafe, exothermic or endothermic? Explain.

8. Find the best possible solution to the following equations.
   (i) $3x + 4y = 8$
   $2x + 5y = 7$
   $2x + 6y = 9$

   (ii) $3x + 4y + 5z = 11$
   $4x + 5y + 8z = 19$
   $x + 2y + 4z = 8$
   $5x + 6y + 10z = 23$

9. What does steady state mean? Write the general mass balance equation. Identify the various terms of the mass balance equation for a first-order reaction occurring in a CSTR, under isothermal conditions.

5

Dimensionless Analysis and Scale-up
Another Illustration of How Physics and Mathematics can be Combined

Introduction

The development of a technology or a process usually starts in a laboratory. In the development of a process involving a reaction the engineer or scientist is interested in determining whether a reaction is feasible or not. This information can be found by studying the free energy change of a reaction. If it were to be negative, then the reaction can occur from the thermodynamics point of view. This, however, only tells us that the reaction is possible. If the free energy change were to be positive, then the reaction can definitely not occur and the scientist should abandon all possible attempts to pursue the reaction using those reactants and should look for an alternative reaction scheme to manufacture the product.

Thermodynamics, however, does not give any information on how fast the rate of the reaction can be. The rate of the reaction can be very slow, in which case it may take a few years to obtain any product. In this situation it is necessary to determine the methods to accelerate the reaction. This could be achieved by adding a catalyst, or by increasing the concentration of the reactants or by increasing the temperature. Hence, optimum conditions have to be determined for the reaction. This is done on a lab scale, where the amount of products manufactured is of the order of a few grams to a few kilograms and the reactors have a low volume of a few litres. After a successful demonstration at the lab scale the idea is scaled up to the commercial scale of production. Here the amount of products manufactured is of the order of hundreds of tons.

Safety Issues in the Scale-up

One of the important responsibilities and challenges of a chemical engineer is the scale-up of a process. The manufacturing steps are first demonstrated at the laboratory scale and it is the primary responsibility of the chemical engineer to come up with a process which will enable high volumes of production. This is now illustrated with a concrete example. Consider a reaction which has been successfully demonstrated in the laboratory as capable of producing 5 moles of a compound in a reactor of size 2 litres (say a beaker or a conical flask). These numbers are hypothetical and are used to illustrate the idea. It is now desired that at the commercial scale 5,000 moles of this compound be produced. You are asked to determine the size of the reactor.

For simplicity, let us assume that the reactor is being designed for a batch operation. One might be tempted to think that a 2,000 litre reactor would serve the purpose. Under this assumption, an increase in amount produced by a factor of 10^3 can be achieved by increasing the reactor volume by the same factor. Here we have assumed that the moles produced depend linearly on the volume. Under this assumption an increase in the amount produced by a factor of 1,000 can be achieved by increasing the reactor volume by the same factor. However, this assumption is frequently invalid since different processes occurring in the system scale with size in different ways. For instance, if we have a homogeneous chemical reaction which is exothermic the heat is generated throughout the volume. The reaction is accompanied by heat generation and heat loss to the ambience. Hence increasing the size of the vessel would result in the heat generation being proportional to the volume, i.e. it increases as a cube of the length scale (radius if the vessel is spherical). However, the surface area across which heat loss to the ambient occurs increases only as the square of the length scale (radius) since it is proportional to the surface area. Hence increasing the volume of the vessel implies that the rate of heat generation increases much more than the rate of heat loss. Consequently, this would result in a higher temperature prevailing in the reactor at the commercial (larger) scale as opposed to the lab (smaller) scale. This has implications on the reaction rate which is dependent nonlinearly (exponentially) on temperature through the Arrhenius temperature dependency. The point being made here is in the larger reactor as the reaction progresses the temperature prevailing would be different from that in a smaller reactor. The bigger reactor may have to be designed to ensure that the temperature is the same as that in the lab scale or there could be undesirable side reactions giving unwanted products reducing the reaction efficiency. If the objective is to keep the temperature same it is necessary to provide additional provisions in the larger reactor to remove the extra heat being generated.

A second problem that can arise is maintaining spatially uniform conditions in the larger reactor. In a smaller vessel it may be easier to have spatial uniformity. The spatial variations in a larger reactor may result in the performance of the larger reactor being different from what is expected. Hence, during the scale-up these issues have to be kept in mind and addressed.

The scale-up issues pose challenges for several systems, in particular exothermic reactions which exhibit runaway behaviour. What does *runaway* mean? In an exothermic reaction as the reaction proceeds, heat is liberated. This causes a further increase in the temperature which in turn makes the reaction proceed faster. This results in a greater heat release accompanied by a further increase in temperature. Consequently, the temperature rise in the system is very rapid. However, this does not proceed indefinitely as the reactants get consumed. Finally, when no reactants are left the reaction stops. A sudden rapid increase in the temperature of the reactor sustaining an exothermic reaction is called a *runaway behaviour*.

This important feature of the exothermic reaction arises as it is self-sustaining. The behaviour of the system is similar to that of an autocatalytic system where there is a positive feedback effect. In the latter as the reaction propagates a product is formed. This acts as a catalyst and the reaction becomes faster. This in turn results in an increase in the product concentration further accelerating the reaction. The above feature of an autocatalytic and an exothermic reaction is characteristic of an unstable system. Here a disturbance which results in an increase in the temperature or product formed propagates in such a manner that there is a further increase of temperature or product formed. This increase in temperature can cause vaporization of the reactor contents and a consequent increase in the pressure. A possible explosion could occur in the reactor if it is not built to withstand this high pressure. Hence, whenever there are exothermic reactions safety issues have to be addressed specially during the scale-up, since as we have seen the heat generation and heat loss parameters scale differently in a unit when we change the scale.

Lab Scale and Commercial Scale

The demand for a product can be a few hundred tons per year. Under these circumstances, this large volume can be produced by carrying out the reaction in tens of thousands of the small reactors in the laboratory. This would mean that the production is carried out in parallel. This would be a very inefficient way of doing things as it would require a large equipment inventory, it would be highly labour intensive, etc.

It is for this reason that most plants that we see in the industry are large in size. Here the plant operation becomes economical when it has a large capacity. This is the so-called *economies of scale* where the utilization of energy, raw materials, etc. on a per unit product formed basis is very efficient. The efficiency increases with the integration of different units and streams so that heat liberated in one unit can be used in another unit.

How does one go from the lab-scale production to the commercial scale? As seen above there are several challenges (mixing, rate of heat generation, and heat loss) when the size of units is increased. New physical phenomena arise when the scale is increased and these have to be accounted for. Hence increasing the size is usually not done in one step. The scale of the plant operation or the size of the plant is not increased by three orders of magnitude, for instance. The system behaviour in an intermediate scale called the *pilot plant* is first analysed. This has a production capacity which is in between the lab scale and the commercial scale. For instance, it could be a plant with a capacity of tens of tons. The process is first extended from the lab scale to the pilot-plant scale. There is a significant amount of learning or knowledge which is gained at this stage. One new effect could be poor mixing which can be present as the size increases. These new effects have to be accounted for. This is achieved at a moderate cost, since the costs for building the pilot plant and running it are much lower than that of building and running a commercial plant. This knowledge is then used to scale up further to the commercial scale. Sometimes pilot-plant level tests may be done at two intermediate scales before going for commercial production.

Let us see some of the challenges involved in the scaling-up. For instance, when we consider the energy balance equation we see that the heat loss term is proportional to the surface area of the reactor and this scales as a square of the length scale. The heat generation term is proportional to the cube of the length scale. The surface area per unit volume, hence, scales as the reciprocal of the length scale. To increase the production by a factor of 27, then it is not enough to just increase the radius by a factor of 3. This may not work as seen earlier. The heat produced in the smaller scale may be dissipated to the ambience since the surface area per unit volume is high while in the larger scale the surface area per unit volume is low and, hence, the heat loss is low and we would have an increase in the temperature inside the reactor. The operating conditions (initial concentrations of the reactants, initial temperature, etc.) which were safe for the pilot plant may be unsafe for the commercial scale reactor. These challenges make it important for us to obtain a good understanding of all the processes (reaction, mixing, and heat transfer, etc.) which are taking place in the system so that the scale-up can be done in a reliable way with minimum scope for any error.

To illustrate this further in a semi-quantitative manner the energy balance for a batch reactor sustaining a first-order reaction is considered. A batch reactor is a dynamic system. Here as the reaction progresses, heat is liberated if the reaction is exothermic and the temperature increases as a function of time. Hence, the accumulation term cannot be set to zero since we do not have a steady state. The energy balance equation is

$$\text{Accumulation} = \text{In} - \text{Out} + \text{Generation} \tag{5.1}$$

For a batch system there is no inflow or outflow of reactants or products across the reactor. The generation term represents the heat lost by the reactor to the ambient if the reactor is not insulated and the heat generated by the exothermic reaction. This term is high when the reaction rate and the heat of reaction are high. When this is higher than the outflow term we have a net accumulation of heat resulting in an increase in the temperature with time. If the outflow term is more than the generation term the accumulation term is negative and this results in a decrease in temperature as a function of time.

Let us look at the mathematical expressions for these terms now, i.e. how can they be quantified. The outflow term arises because of the heat loss from the reactor to the ambient. It is proportional to the temperature difference between the reactor and the ambient and also varies directly with the surface area available for the heat transfer. Thus, the rate of outflow of heat is

$$q_{\text{out}} = hA(T_r - T_a) \tag{5.2}$$

Here the parameter h represents a heat transfer coefficient and is the rate of heat loss per unit area when there is a unit temperature difference. This would depend on the conditions prevailing inside and outside the reactor, i.e. the flow rate of air and how well the liquid is stirred, etc. The higher the air velocity in the ambient, the more will be h.

The generation term is given by $(-\Delta H)VCk_0 e^{-(E/RT)}$. This represents the heat generated by the exothermicity of the reaction in J/s. It is seen that the heat generation term is dependent on both the heat of reaction and the rate of the reaction. If a reaction is highly exothermic with a very low rate there may not be enough heat generated and, hence, we are likely to see very little change in the temperature. In other words, any heat generated will be lost to the environment since the outflow is more than the generation. On the other hand, if the outflow term is drastically less than the generation term, then the temperature would increase drastically.

The outflow term can be altered by changing the area for heat transfer A or by changing h. The latter can be done by increasing the circulation of a coolant fluid or ensuring the ambient fluid flows at a higher rate than what is prevailing or lowering the temperature of the coolant fluid (if the reaction is exothermic).

Should the temperature be sufficiently low the reaction may not progress and this would also be undesirable. If the temperature were to be very high, then there could be a runaway reaction and this would be undesirable, too. It is, hence, necessary to control the temperature at a reasonable level to ensure the reaction occurs smoothly and safely. This has to be kept in mind during the scale-up of an exothermic reaction.

Dimensionless Analysis: Dimensionless Numbers

In the earlier section the dependence of the system behaviour on the size or scale of the system has been explained. The concept of dimensionless analysis and dimensionless numbers helps us understand the scale invariance in the system behaviour. This also helps us understand the behaviour in more general terms. This is now illustrated with an example from fluid mechanics.

Ships moving through water and aeroplanes moving through air have to overcome a *drag force*. The origin of this force lies in the viscous resistance offered by the fluid through which it is moving. *Viscosity*, an intrinsic property of fluids, is the equivalent of friction. Hence, the viscous resistance force always acts in the direction opposite to the ship or aeroplane motion. One of the aims of the engineer is to design the aircraft or ship, i.e. determine the shape, such that it minimizes the drag force or resistance to its motion. The streamlined bodies of the aircraft ensure that the drag resistance is low. To determine the drag force on an actual design of a ship, it is necessary that we build the ship and determine the drag force on it for various velocities. This can be done experimentally but would need fabricating the entire ship. If now a different design has to be tested for the drag force the ship has to be re-built according to this design and the drag force measured experimentally. This would, hence, be a practically impossible approach since each new design has to be fabricated to the full scale and tested. This would be a costly proposition.

One methodology to avoid this problem is to design and fabricate models which are smaller in size, say by 1:10 or 1:50 and perform experiments on them. By this we mean the model dimension is 1:10 or 1:50 of the actual system dimensions. This model must be "similar" to the original full-scale system. One necessary condition to ensure similarity is that all linear dimensions must scale at the same ratio. More details of this can be found in (White 2010). The drag force on these models is calculated in a wind tunnel or a water tunnel in a laboratory. Here the model is kept stationary and the fluid moves around it. In the actual situation the body is in motion over a relatively stationary fluid. These two situations are identical as only the relative velocity determines the drag force on the body.

Drag forces are experimentally measured on the small-scale model. The advantage now is that the small-scale model can be fabricated with ease and several designs, and alternatives can be tested without significant implications on costs. It is now of interest to see how this information can be used to estimate the drag force on the full-scale object. For this, we identify properties or groups of variables which are scale invariant. This group of variables has the same value at both scales. The drag force on the prototype is used to determine the power of the engine required to ensure that we have enough thrust to overcome it.

In the field of chemical engineering, drag forces play an important role. Sedimentation tanks are used to separate solids in a slurry. This is a common operation in wastewater treatment. Here gravity is allowed to do the work of separating the solids from the liquid. The heavier particles settle down under the action of gravity first. To understand this, let us consider how a single particle settles down in a quiescent liquid. The behaviour of the particle is determined by the forces acting on the particle and these are the gravity force, buoyancy force, and the drag force. The first one acts downwards and the other two act upwards for a downward settling particle. For a short period of time when the particle enters the water there is a net force acting downwards on the particle. After this as the particle moves further downwards the forces balance each other and the net force is zero. This ensures the particle settles at a constant terminal velocity. This settling velocity would be a function of the size of the particle, properties of the solid, and the fluid. One way to estimate this is to carry out experiments on the system of interest. This would mean doing experiments to obtain terminal velocity for different combination of fluids, solids, etc. The dependence of terminal velocity on one of the variables, say, size can be depicted graphically. This would be valid only for the particular liquid-solid combination. It is, hence, desirable to have a method by which the drag force or terminal velocity can be predicted for any given system, i.e. any particle size or liquid-solid combination in a universal manner. The problem in an actual sedimentation tank is more complex as the different particles interfere with each other during the settling process.

The principle on which the scale-up is based is called the *dimensionless analysis*. Dimensionless analysis is similar to the concept of "*similarity of triangles*" which you have seen in a course in geometry in your school. Here, the ratio of two sides of a bigger triangle is the same as the ratio of the corresponding sides of a smaller but similar triangle. If two sides of a triangle are known and a corresponding side of the bigger triangle is known the other corresponding side of the bigger triangle can be found. In dimensionless analysis we match the ratio of variables such as forces in the two different objects.

First, physical and mathematical foundations of the principles are discussed and then they are illustrated with examples. The scale-up of a system is concerned with the methods to estimate the information on the actual scale using data collected on a smaller model.

The primary idea in dimensionless analysis is to understand and identify dimensionless groups which determine system behaviour at various length and time scales. This approach is useful as long as no new physical interaction or effect arises when we go from one scale to another. In dimensionless analysis we use ideas from physics and mathematics and integrate them to predict the behaviour of large-scale systems after analysing the behaviour of a small-scale system.

This concept is first illustrated by considering an example: the pressure drop in a pipe (a problem close to the chemical engineer's heart). This problem is easier to explain as the physics is simpler than that on the drag force on objects. The drop in pressure arises since the liquid has to overcome the viscous resistance at the walls of the tube. The physical origin of the pressure drop is, hence, the same as that of the drag force on a moving body. The pressure drop across a horizontal pipe would depend on several variables. The list of these variables is determined by the physics of the process. Thus, we expect the pressure drop to be a function of fluid density (ρ), fluid viscosity (μ), velocity (v), diameter (D), and length (L). This can be written abstractly as

$$\Delta P = f(\rho, \mu, v, D, L)$$

In writing this equation information from the physics of the problem has been used. From our understanding of the physics of the system no other variable can influence the pressure drop. The only drawback is the exact form of the function f is not known. For a particular fluid flowing through a specific pipe (i.e. for a fluid with a fixed ρ, μ, D, L), the dependency of the pressure drop on velocity $\Delta P = f(v)$ can be found experimentally. This can be graphically depicted. However, this information cannot help us estimate the pressure drop at different velocities if we change the fluid or the pipe. For a different fluid flowing through a different pipe we have to carry out experiments again to determine the dependency of pressure drop on velocity. The functional dependency determined is likely to be the same qualitatively but can be quantitatively different.

Pipe flow experiments can be performed where a fluid is pumped through a pipe of known length L and diameter D. The average velocity of the fluid can be estimated from the volumetric flow rate which can be experimentally determined. The properties of the fluid ρ (density) and μ (viscosity) are known when the fluid is fixed. The pressure drop can be measured by using a manometer and its dependence on v can be shown. A typical plot of this dependency is shown in Figure 5.1. This would, however, be for a fixed value of all the other

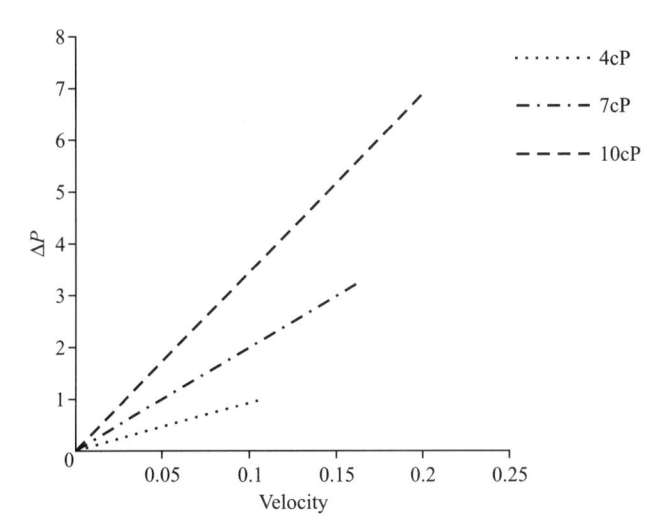

Figure 5.1 Pressure drop dependency on velocity through a pipe for laminar flow.

114 *Introduction to Chemical Engineering*

parameters ρ, μ, L, D. The different curves represent the dependency of pressure drop on velocity for three typical values of viscosity. The graph, hence, has limited validity. If we were to use a different pipe (different L or D) or a different fluid (different ρ or μ), this graph cannot be used to determine ΔP for different velocities. Thus, we need several thousands or millions of experiments or graphs to take into account the millions of different possible combinations of the various properties. The practising engineer interested in determining the pressure drop for his system would have to use these graphs. Or, he would have to do experiments on his system and determine the pressure drop. This is not a very practical way of getting the information of pressure drop.

It would, hence, be desirable if a universal function could be obtained which can be used to determine the pressure drop in a pipe for any fluid. In other words, is it possible to seek a compact representation of the data so that one can depict all the information of all the experiments in one graph?

Writing the equation for pressure drop as a function of the other variables implies that each term has the dimensions of pressure drop (Pa/m). This equation can be recast into another form such that each term is dimensionless. Here we rewrite the equation such that each term contains a product of variables raised to a power. The question that arises is: can we estimate a combination of ΔP (dependent variable) and certain independent variables as a universal function of a combination of the independent variables (ρ, μ, v, D, L)? The existence of such a function can be established by using the fact that in an equation where several terms are present, each term must have the same units or dimensions. This is called the *principle of dimensional homogeneity*. As a result of this, any equation can be recast such that each term has no dimensions by an appropriate manipulation. This conversion to a dimensionless form yields dimensionless numbers (groups of variables which are dimensionless). These numbers are groups of variables which do not have any units. Converting an equation to the dimensionless form results in a relation with fewer dimensionless variables as compared to the original number of variables which had dimensions. In the example of the pressure drop problem the total number of dimensional variables is six, but when converted to dimensionless variables the number reduces to three. This allows us to represent information in a compact form. Let us see how we can arrive at these dimensionless numbers.

Each of the variables involved has specific dimensions. We write the dimensions of these variables in the form of a vector.

The fundamental units in which all variables are expressed are mass (M), length (L) and the time (T). Density has dimensions of kg/m^3 and can be represented as a vector [1 − 3 0], where the first element represents the power to which mass is raised in the dimension of the variable, the second the power to which length is raised, and the third the power to which time is raised. Thus, each variable has a vector representation. Velocity, for instance, can be represented as [0 1 −1].

This leads us to represent the system in terms of the variable matrix where the units of each variable are written in a vectorial form. Our system has six variables and these are written as

$$\begin{array}{c|ccc} & M & L & T \\ \hline \rho & 1 & -3 & 0 \\ v & 0 & 1 & -1 \\ D & 0 & 1 & 0 \\ L & 0 & 1 & 0 \\ \mu & 1 & -1 & -1 \\ P & 1 & -1 & -2 \end{array}$$

The matrix is of size (6 × 3), i.e. it has 6 rows (corresponding to the number of variables) and 3 columns (corresponding to the fundamental quantities or units of all variables). The rank of a matrix represents the number of independent rows or columns. For a matrix the row rank and the column rank are identical (the number of independent rows equals the number of independent columns). The rank of this matrix is 3. If the rank is three, then there are three independent rows and every other row can be expressed as a linear combination of the three independent rows. The three independent rows can be chosen in several ways.

It is easy to see that the first three rows are independent. They can, hence, be chosen as a basis, i.e. every row in the matrix which has three elements can be written as a linear combination of the first three row vectors. This helps us physically find a dimensionless group as a product of variables where the first three variables and another variable are chosen. Consider the case when the fourth variable is the length of the pipe. The row vector corresponding to L, [0 1 0] is written as a linear combination of the rows corresponding to ρ, V and D. In Eq. (5.3a), we write the vector for L in terms of the three basic variables. In Eqs. (5.3b) and (5.3c), we write the variable μ and ΔP in terms of the three basic variables.

$$[0 \quad 1 \quad 0] = a_1[1 \quad -3 \quad 0] + b_1[0 \quad 1 \quad -1] + c_1[0 \quad 1 \quad 0] \quad (5.3a)$$

$$[1 \quad -1 \quad -1] = a_2[1 \quad -3 \quad 0] + b_2[0 \quad 1 \quad -1] + c_2[0 \quad 1 \quad 0] \quad (5.3b)$$

$$[1 \quad -1 \quad -2] = a_3[1 \quad -3 \quad 0] + b_3[0 \quad 1 \quad -1] + c_3[0 \quad 1 \quad 0] \quad (5.3c)$$

Each of Eqs. (5.3a to 5.3c) gives rise to a system of three equations which can be solved to determine:

$$b_1 = 1$$
$$a_2 = b_2 = c_2 = 1$$
$$a_3 = 1$$
$$b_3 = 2$$

and

$$a_1 = c_1 = c_3 = 0$$

116 Introduction to Chemical Engineering

This implies that the following variables are dimensionless:
$$\frac{L}{D}, \frac{\rho V D}{\mu}, \frac{\Delta P}{\rho V^2}$$

The choice of the dimensionless variables is not unique and is dependent on the choice of the basis variables (variables which are chosen to be independent in the matrix). Since any three variables can be chosen as an independent set we can form different dimensionless groups. However, the number of dimensionless variables which govern the system behaviour is unique. This is given by the difference in the number of variables and the rank of the matrix formed.

For instance, if μ, V, D are chosen as independent variables, then the three dimensionless groups which arise are $\frac{L}{D}, \frac{DV\rho}{\mu}, \frac{\Delta P D^2}{\mu}$. An alternative but equivalent approach to determining these dimensionless groups is to seek each of the non-basis variables as a product of the three basis variables raised to an exponent. We exploit the fact that the two sides of the equation must have the same units.

Now let us illustrate this using viscosity, the fourth variable and density, velocity and diameter as the basis variables.

$$\mu = \rho^a V^b D^c$$
$$M^1 L^{-1} T^{-1} = M^a L^{-3a} \cdot L^b T^{-b} L^c \tag{5.4}$$

Equating the powers, we get

$$1 = a$$
$$-1 = -3a + b + c$$
$$-1 = -b$$

This is the same as Eq. (5.3b). This process can be repeated by considering the pressure drop and the length of the pipe.

If we were to view the function as one where all the terms are dimensionless, we can interpret it as being an equation which relates dimensionless variables. The function in terms of these variables which we have now identified can then be recast as

$$\frac{\Delta P}{\rho V^2} = f\left(\frac{L}{D}, \frac{\rho V D}{\mu}\right) \tag{5.5}$$

This implies that the dimensionless group on the left is a function of the dimensionless groups on the right. Increasing the length of the pipe increases the pressure drop proportionally. This has been experimentally observed as long as the shape of the velocity profile does not change when the length is increased. So we expect the dependency on L to be linear, and we write this as

$$\frac{\Delta P}{\rho V^2} = f\left(\frac{\rho V D}{\mu}\right) \cdot \frac{L}{D} \tag{5.6}$$

The dimensionless group $\rho VD/\mu$ is called the *Reynolds number*, Re. This has a physical meaning. It is the ratio of inertial to viscous forces. At low Re the viscous forces dominate and the flow is called the *laminar*. As the inertial forces dominate Re increases and the flow becomes turbulent.

We have

$$\frac{\Delta P}{\rho V^2} = f(\text{Re}) \cdot \frac{L}{D} \tag{5.7}$$

The function which captures the dependency on Re is called the *friction factor*. The exact form of f can be determined by experiments. To do this we would have to vary the velocity in a pipe and determine pressure drop across the pipe for different velocities. For a pipe of given length a plot of $\Delta P/\rho V^2$ as a function of Re would graphically represent the function f. This is called the *Moody diagram* (Figure 5.2) and should be universally valid for all fluids and pipes as long as the pressure drop in a pipe does not depend on any other variable.

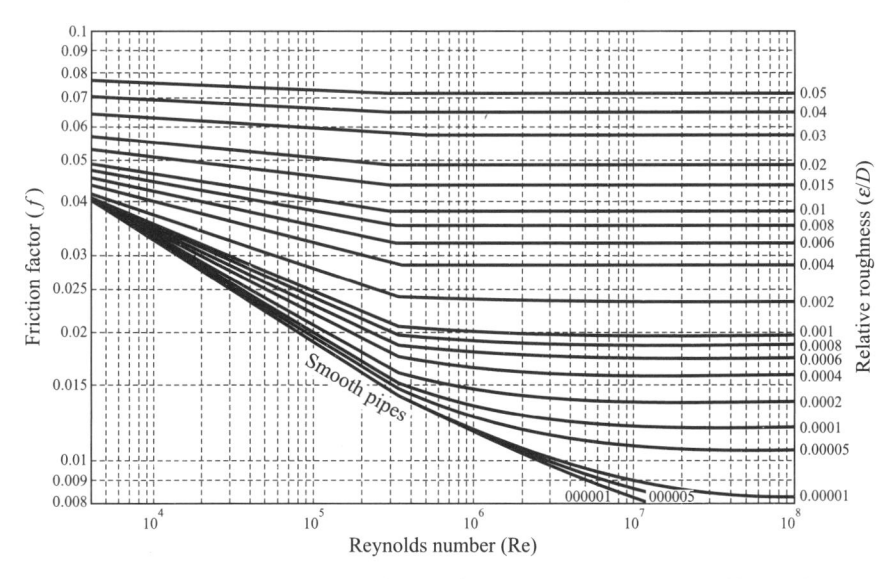

Figure 5.2 Moody diagram showing friction factor dependency on Reynolds number (Re).

Suppose the material of construction of pipe is changed. Pressure drops in pipes of the same size for the same flow rates are different when the material of construction is changed. This implies a new physical parameter comes into the picture. In terms of dimensionless analysis this means there is an additional parameter which comes in to characterize the pressure drop, i.e. the surface roughness. This depends on the material of the pipe and is characterized by ε.

An additional dimensionless parameter ε/D now enters into the picture. We now have a family of curves in the turbulent region each characterized by a different value of ε/D. The Moody diagram in fluid mechanics contains this family of curves for different ε/D. This family of curves is generated experimentally. Alternatively, the experimental results can be fitted to functional forms and this can be used to compute the friction factor. In Figure 5.2, we obtain the dependency using the *Nikuradse equation* which is a function which captures the dependency of f on Re and the roughness.

This diagram can be used to determine ΔP across a pipe through which a fluid flows. The advantage of using this approach is that the function f is now unique and valid for all liquids as well as all pipe sizes. This has been made possible since we now plot the scaled or dimensionless variables instead of the actual physical variables. Experiments can, hence, be performed on a fixed length of a pipe and the functional form of f on Re can be graphically determined by varying velocity alone. The function can then be used to predict the friction factor and hence pressure drop for a larger pipe with a different fluid which may be of interest. For the larger pipe, we would determine Re and then compute f. From f, the pressure drop would be determined. Thus, dimensional analysis greatly reduces the number of experiments to determine the pressure drop in a pipe by a compact representation.

This approach depicts the integration of physics and mathematics. We use physics in enumerating the variables which are dependent on each other. If our understanding of the physical situation is complete, the list of variables is complete. If the physical behaviour of the system depends on some other variable it has to be added to the list of independent variables. Mathematics tells us that an equation must be dimensionally homogeneous, i.e. every term must have the same dimensions or units. When we render an equation dimensionless all terms are without dimensions. Hence, we look for variables or groups of variables which are dimensionless. Mathematics tells us that the dimensionless groups which are identified are not unique. We can obtain different dimensionless groups by choosing different variables as the independent basis but the number of dimensionless groups will always be the same. Thus, we have a scientific approach which is based on physics and mathematics and this gives a scientific rigour and basis to our analysis. The experimental data helps us establish the functional dependency, at least graphically and hence the accuracy of results is guaranteed. This approach also helps us extrapolate information over a wide spectrum of length scales.

Obtaining the Friction Factor

One way to determine the function f is using experiments. The question one can ask is: Is it possible to estimate the function f from a purely theoretical approach? In some cases it is possible. The nature of the flow depends on the

Reynolds number. When Re < 2,100 the flow is *laminar* and for Re > 2,100 the flow is *turbulent*. At low velocities when the flow is laminar, the fluid has only an axial velocity component. The other two velocity components along the cross-sectional area in the radial and θ direction (of a cylindrical pipe) vanish. It can be shown that the axial velocity profile is parabolic for laminar flow.

In turbulent flow all three velocity components are present and are time dependent and the flow is unsteady. It is not possible to obtain an analytical solution for the velocity field.

For the laminar flow it can be proven that $f = 64/\text{Re}$. This is an exact theoretical prediction for the friction factor which can be experimentally verified and is obtained from an application of the law of conservation of momentum. However, the validity of this analytical relation is only for Re < 2,100, when the flow is laminar. This limit has been experimentally identified. Let us see how this can be proven theoretically.

Consider the fully developed laminar flow of a liquid in a cylindrical pipe. It can be shown that the velocity profile (axial component of velocity) is parabolic.

$$\left[v(z) = v_{max}\left(1 - \left(\frac{r}{R}\right)^2\right) \right] \quad (0 \leq r \leq R) \tag{5.8}$$

This profile implies that the velocity is zero at the walls (where $r = R$) and is a maximum at the centre (where $r = 0$). This velocity profile is schematically shown in Figure 5.3. The other two velocity components are zero. The velocity profile is the same at $z = 0$ and $z = L$ as the flow is fully developed.

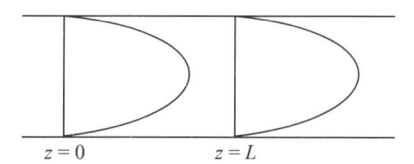

Figure 5.3 Fully developed parabolic velocity profile. Fully developed implies no change in the velocity in the direction of flow.

Consider the control volume as $0 < z < L$, $0 < r < R$. The flow is fully developed, i.e. the velocity profile is independent of z the axial coordinate. The conservation of mass is, hence, automatically satisfied: here the mass coming in at $z = 0$ equals the mass leaving at $z = L$. We apply the conservation of momentum.

Accumulation = In − Out + Generation

As the flow is steady there is no accumulation of momentum in the chosen control volume and this term is *zero*.

Consider the annular element of thickness dr, varying from r to $r + dr$. Its area is $(2\pi r dr)$. The mass flow rate through this is $\rho v_z 2\pi r dr$, and the associated momentum entering is $\rho v_z^2 2\pi r dr$.

The rate at which momentum enters the control volume at $z = 0$ is
$$\int_0^R \rho v_z^2 2\pi r dr.$$

This is the same as the momentum leaving the system as the velocity profile has not changed when the fluid moves from $z = 0$ to $z = L$. Consequently, the rate at which momentum enters and leaves the system is equal.

This implies that the rate of generation of momentum in the control mass associated with the control volume is 0.

From Newton's law the rate of generation of momentum in the control mass occupying the control volume is given by the forces acting on the system. Hence, we come to the conclusion that the sum of forces, i.e. net force acting on the control volume is zero. Let us now see what the different forces are which act on the mass in the control volume. There are pressure forces acting at $z = 0$, $z = L$ and the shear forces acting along $r = R$.

The pressure forces acting on the control volume in the z-direction is

$$(P_0 - P_L)\pi R^2$$

where P_0, P_L are pressure acting at $z = 0$, L.

The shear force is $\tau_{rz} 2\pi R L$.

Assuming the fluid is Newtonian, i.e. the relationship between the shear stress and strain rate is linear, we obtain

$$= \mu \frac{\partial v_z}{\partial r}\Big|_{r=R} 2\pi R L$$

$$= -\mu v_{max} \frac{2}{R}(2\pi R L) \quad (5.9)$$

The negative sign shows the direction of the force. It can be shown for the parabolic profile that the average velocity is half of the maximum velocity. This yields

$$(P_0 - P_L)\pi R^2 = 8\mu v_{avg} \pi L$$

In the above equation each term has the same dimensions or units. This can be recast so that the terms on both sides are dimensionless.

$$\frac{(P_0 - P_L)}{\frac{\rho v_{avg}^2}{2}} = \frac{8\mu L}{\rho \frac{v_{avg}}{2} R^2} = \frac{64\mu}{\rho v_{avg} D}\left(\frac{L}{D}\right)$$

$$\left(\frac{(P_0 - P_L)}{\frac{\rho v_{avg}^2}{2}}\right) = \frac{64}{Re}\frac{L}{D} \quad (5.10)$$

The laminar fully developed flow (parabolic velocity profile) can be experimentally realized for Re < 2,100. Comparing the above expression with Eq. (5.7) results, it is seen that the friction factor f is 64/Re. This friction factor is called the *Darcy friction factor*. The Mooody diagram shows Fanning friction factor which is 16/Re. The exact functional dependency, hence, has been obtained analytically by applying the conservation of momentum. We can experimentally verify this relationship as long as the flow is laminar.

For Re > 2,100 the flow is turbulent. Consequently, the velocity profile is unsteady and the above relationship breaks down. However, our knowledge of physics indicates that the friction factor (f) depends only on Re. This dependency can be found experimentally and plotted. This has been done in Figure 5.2, which can now be used directly for all Re (any pipe size and velocity).

For the turbulent flow regime an exact analytical form cannot be obtained. Here we carry out experiments and determine f for different Re. This is plotted as a smooth curve. Alternately, a mathematical expression can be fitted to the curve and an empirical relation obtained. This can be used to estimate the pressure drop. This function obtained is universal since f does not depend on any other parameters. The f for the laminar region can be obtained from theory (conservation of momentum) and/or experiments while for the turbulent region it is obtained only from experiments. The power of dimensionless analysis lies in the fact that it tells us that the dependency on Re is unique, and no other dimensionless variable is required to calculate the pressure drop. This helps us estimate the pressure drop with confidence.

This relation also demonstrates the principles of scale-up and how dimensionless groups are used in scale-up. For instance, experiments can be carried out in the laboratory and pressure drops can be obtained on a small scale set-up for the entire range of Re. How can this be used for scale-up? The experimental results can be plotted in terms of dimensionless variable as shown in Figure 5.2. For a given operating condition characterized by Re, the figure can be used to find the friction factor f and then the pressure drop ΔP. This same graph can be used to find ΔP for a large scale system with the same Re (e.g. larger diameter but appropriate lower velocity). Here, as Re is the same, the friction factor f will be the same for the larger system, however the pressure drop across it will be different. The dimensionless approach helps us perform the scale-up, as long as when we increase the length scale no new physics has to be added, i.e. the list of parameters does not change. Here as the Re is increased, the flow changes from laminar to turbulent and the functional dependency becomes different but the independent variables remain unchanged.

To compute the pressure drop, we use the properties and average velocity and find Re. We obtain f from the graph and multiply that by $\dfrac{L}{D}\dfrac{\rho v^2}{2}$ to obtain ΔP. This can be done for any flow rate, fluid, length and the diameter of a pipe.

122 Introduction to Chemical Engineering

Here, we now have only one graph from which we can obtain ΔP. By using dimensionless numbers we have been able to capture the information present in the thousands of graphs discussed earlier and present them in one graph.

This example shows that a rigorous basis of science that is physics and mathematics along with realistic experimental data allows us to solve problems of engineering interest elegantly. The dimensionless numbers which are obtained are chosen to have a physical significance.

Application of Dimensionless Analysis to a Reactor, i.e. A CSTR

We now show an example of how the exact dimensionless relationship or dependence between a dependent variable and an independent variable can be obtained from first principles. In this example the law of conservation of mass is applied to a reactor. The governing equation is modified by algebraic manipulations and recast into dimensionless form. The relationship we seek arises naturally from the solution.

Consider the reaction occurring in a continuously stirred reactor. In this reactor the reactants are fed into the reactor which can be visualized as a vessel. There is a stirrer which keeps the contents of the reactor at a uniform composition and temperature. The concentration in the reactor and of the reactor exit are assumed to be the same.

Under these conditions the concentration of the reactant in the exit stream can be expected to be a function of the feed concentration C_f, the reactor volume V, the flow rate of the reactant stream q, and the reaction rate constant k. Thus, we can write

$$C = f(C_f, k, q, V)$$

There are a total of five variables and, hence, we expect a relation between two dimensionless groups. The variable matrix can be set up and the dimensionless groups can be determined as C/C_f and Vk/q for a first-order reaction. Thus, we can expect that C/C_f is dependent on Vk/q. What the exact nature of the dependency is can be obtained from experiments or by using the conservation of mass which we have seen earlier. For a reactor at steady state sustaining a first-order reaction the conservation of mass gives

Accumulation = In − Out + Generation

At steady state, we have for the reactant concentration C,

$$0 = qC_f - qC - VkC \tag{5.11}$$

Our objective is to find C. Clearly $C = f(q, V, k, C_f)$ and is given as

$$C = \frac{qC_f}{q + Vk} \tag{5.12}$$

We can make these dimensionless. For this, we define conversion $X = 1 - \dfrac{C}{C_f}$.
We obtain an expression for conversion as

$$X = \frac{Da}{1+Da} \quad (5.13)$$

where $Da = Vk/q$ is a dimensionless group called the *Damkohler number* measuring the ratio of two time constants the ratio of the residence time (V/q) to the reaction time ($1/k$). The reaction time represents the time required for the concentration level to drop down by a fixed percentage from that prevailing in the feed. The dependency of X on Da is depicted in Figure 5.4. This is generated using a MATLAB program (MATPROG 3) in the appendix.

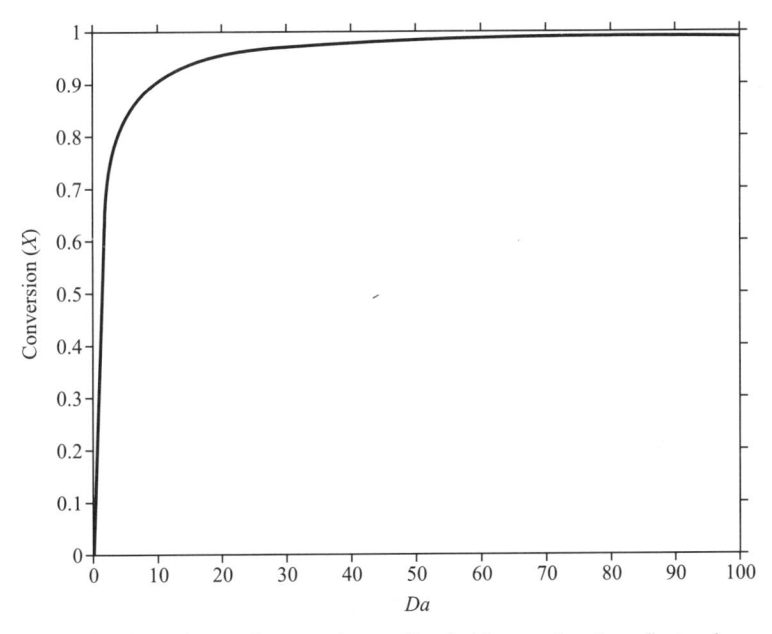

Figure 5.4 The dependency of conversion on Damkohler number for a first-order reaction.

Clearly, at low Da the reactants do not spend enough time in the reactor and the conversion is low. As the flow rate is reduced, Da increases and the conversion rises.

By recasting in dimensionless form, we obtain only one curve depicting $X = f(Da)$. Thus, we can determine the volume of a reactor from the Damkohler number, Da, which would give us a desired conversion level. When we write $C = f(q, V, K, C_f)$, the dependence of C on q can be depicted only for a fixed V, k, C_f, whereas the conversion X dependency on Da is universal. If we plot the dependency of C_A on q, we get a family of curves for different reactor volumes for a fixed k, C_f. This is obtained from Eq. (5.12). The MATLAB program (MATPROG 4) in the appendix generates Figure 5.5.

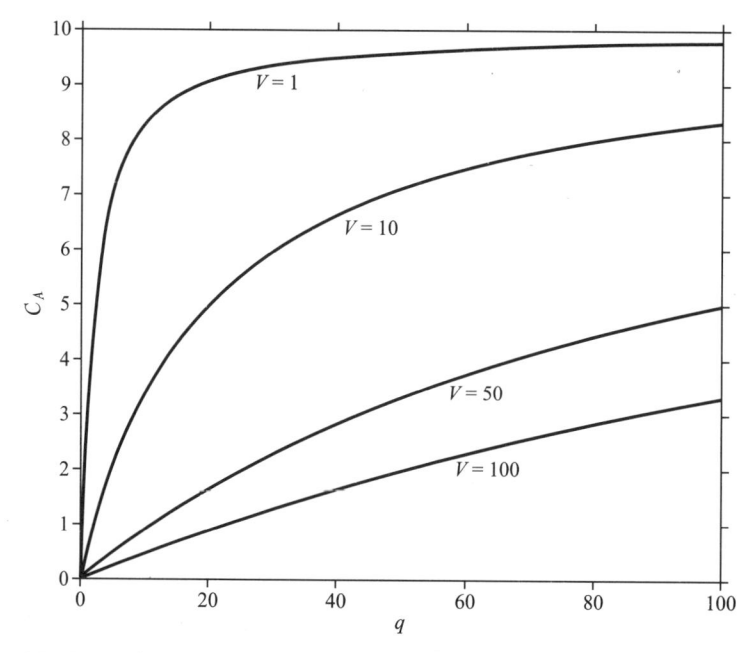

Figure 5.5 Dependency of reactor concentration on flow rate for different reactor volumes.

For a first-order reaction in terms of dimensionless variables $X = f(Da)$ and this representation is unique. This functional dependency of conversion on Da is obtained assuming the reactor is well mixed. The dependency would be different if the mixing inside the reactor is different. While perfect mixing as assumed in the CSTR model is easy to achieve in a small scale it may not be possible to have well-mixed conditions in a large reactor. Hence, the prediction using Eq. (5.13) may be invalid when we scale up to a larger size.

Conclusions

Dimensionless variables help in compact representation of data. Fundamental laws can be used to obtain the relationship between dimensionless variables as seen in the examples of the CSTR and laminar flow. In other cases, a combination of experiments and theory has to be used.

Dimensionless analysis is based on scientific rigour. It helps us understand all variables on which a parameter depends and this dependency is recast in a compact form. In some cases as in the laminar flow in a pipe the dependency can be found analytically. However, in other cases, as in turbulent flows the exact functional dependency has to be found from experiments. However, the strength of dimensionless analysis lies in the fact that the number of variables on which a parameter depends can be identified. This two-pronged approach where we use scientific rigour to identify variables and experiments to get the functional dependency is an example of semi-empirical approach. We will see further examples of this approach in Chapter 6.

Exercises

1. Derive the dimensionless groups which determine the performance of an nth order reaction in a CSTR. Find the exact nature of the dependency from the mass balance equation.

2. Form a dimensionless group with a surface tension, velocity, diameter, and density. These variables arise in flow through small channels when gravitational effects are negligible.

3. Find the dimensionless groups on which the power required for a stirrer in a CSTR depends.

4. Using dimensionless analysis, find the dimensionless groups on which the drag force on a ball depends.

5. A hot spherical body is losing heat to ambient air flowing around it. The heat transfer coefficient (Wm^{-2}/K) is a function of the thermal conductivity, specific heat, viscosity, density, and velocity of ambient fluid and characteristic dimension of the solid.
   (a) Identify the number of independent units in the variables.
   (b) How many dimensionless numbers do you expect? Are these numbers unique?
   (c) In the literature, check what the Nusselt number depends on. Can you reconcile your results to the expressions in the literature?

6. Let us consider the mass transfer analog to the above problem, i.e. a spherical solid body, say, naphthalene undergoing sublimation in an air stream. The mass transfer coefficient (m/s) is a function of diffusivity, viscosity, density, velocity and characteristic dimension of the body. Find the number of independent units of the above variables. How many dimensionless numbers do you expect? Find these numbers. As time progresses the spherical body will lose mass. The size of the sphere shrinks. The problem is now unsteady. Will this influence the dimensionless analysis?

7. Explain how will you determine the velocity in a pipe given the pressure drop across it?

   (**Hint:** The process is iterative now.)

8. In enumerating the variables on which a parameter depends you end up listing a variable which does not have any effect on the parameter. Introducing this additional variable will result in an extra dimensionless group. Can you conjecture how the dependency on this dimensionless group will be reflected in the final equation?

Reference

White, Frank M., *Fluid Mechanics*, 7th ed., McGraw-Hill, New York, 2010.

6

Semi-empirical Approach in Engineering

Departure from Scientific Rigor
Applications in Atmospheric
Pollution and Turbulence

Introduction

In Chapter 5, we saw how for turbulent flows the friction factor is obtained from experiments. It is not possible to analytically characterize and obtain the flow field here and, hence, we have to take recourse to experiments. However, in that chapter we were able to obtain a complete and compact representation of the friction factor graphically since we were assured from the physics that the friction factor depended only on the surface roughness and the Reynolds number. This is an example of a semi-empirical approach to finding a solution.

In this chapter, we will see another approach of semi-empiricism. We start with fundamental laws governing the system using the constitutive relations and include transport processes at the molecular scales. In many cases as turbulent flows we have a situation where it is not possible to solve the system of equations and, hence, certain simplifications are made. These simplifications are based on certain approximations which capture the physical interactions of the system. These introduce certain additional parameters which are determined from experiments. This two-pronged approach (based on fundamental laws and experiments) gives us more confidence in the predictive capability of the models obtained.

Motivation for Semi-empirical Approach

In Chapter 4, we described situations illustrating fundamental laws of physics and their applications to chemical engineering systems. These laws are applicable to all systems. The conservation laws can be written down in detail for any system. Such a representation taking into account all physical details may result in a complex system of equations. The complexity manifests in the form of a large number of variables and a strong coupling between these variables. The information required to solve these may not be available and finding numerical methods to solve these can be challenging. Besides the detailed information predicted by the equations may not be of interest to the engineer who may be satisfied with an overall average prediction. In these situations it is not necessary to describe the system behaviour accurately and completely by a rigorous scientific analysis. The key features of such systems are:

1. Solving these equations could be numerically challenging.
2. In many cases the information about the variables controlling the system and their interactions may not be known.
3. Moreover, the objective and interest may be to obtain an overall macroscopic view of the system. The detailed information coming from the simulations in many cases may not be of interest to the engineer.

In such situations, the governing equations are written down in detail. They are then simplified by making an approximation/assumption which has a physical basis. This eliminates the need to determine the unknown variables and interactions. The approximation introduces new empirical parameters which are determined using experiments. The simplified model containing the empirical parameters is now used to predict the system behaviour.

The two areas where such a situation arises are: (i) in the analysis of turbulent flows, and (ii) in describing dispersion of pollutants in the atmosphere. Both these problems are from Transport Phenomena, the first in momentum transfer and the second in mass transfer.

We first describe the motivation of the problems in these areas, the object of interest is that we would like to determine and see how engineers devise methods to analyse these systems. The analysis cannot be based on a completely rigorous scientific basis combining mathematics and physics. Neither can it be based on purely experimental analysis since this cannot be generalized and the results would be specific to the conditions of the experiment. Here a combination of the two methods is used. The methodology involves beginning with a scientific approach and making some assumptions and approximations. In this process the equations are simplified using physics but some new parameters are introduced. These are determined experimentally. As this approach is based partly on scientific foundation and partly on experimental observations, it is called a *semi-empirical approach*.

Atmospheric Pollution and Dispersion

In the problem of atmospheric pollution discussed in Chapter 3, our aim was to determine the quantitative contributions of different sources to the quality of air, in particular PM_{10} levels at a point. An inverse question can also be asked. Given the emission rates of a particular pollutant from a source in a region can the concentration levels of that pollutant in the region be determined.

If you drive in the neighbourhood of a urea plant you would get a strong odour of ammonia, one of the reactants used. This arises because of leaks of ammonia or from emissions of unreacted gas from a chimney stack. The threshold level of concentration of a chemical above which it is toxic is known. The concentration levels of that chemical in the neighbourhood should be less than this critical threshold value to ensure that the population does not suffer from any adverse health effects.

Let us analyse this problem now from a theoretical viewpoint. Specifically speaking, consider a chimney stack of a factory which emits, let us say NO_x at the rate of Q kg-mol/day. Our interest is in determining the concentration levels at a particular point or in a region of the atmosphere. The concentration at a point is determined by

1. the wind velocity, both magnitude and direction. For instance, the effect of the pollution will be felt only downstream of the source point. While the pollutant is transported primarily in the direction of the wind, there will be some dispersion in the direction perpendicular to the wind as well.
2. the presence of trees and buildings. These in general obstruct the transport of the pollutant and leads to local regions where the concentrations can be high.
3. the stability of the atmosphere. The stability of the atmosphere is determined by the variation of temperature with height above the earth's surface. On a bright sunny day, the ground absorbs the radiation from the sun and the temperature on the ground is much more than that of the air above it. Consequently, the temperature of air decreases as we go up in the atmosphere. In this situation we have a more dense layer of air above a lighter layer of air. This is an inherently unstable situation as the denser air has a tendency to come down and displace the lighter air at the bottom. Under these circumstances, there is good mixing of the layers of air in the vertical direction

due to natural convection. This causes the concentration levels to be uniformly lower in any given region as compared to the case when mixing is poor where there can be local regions of high concentration. The opposite picture prevails on a clear night when the temperature of the ground would be lower than that of the atmosphere. Here a lighter fluid will be on top of the heavier colder fluid. This is a stable situation and, hence, does not facilitate convection in the vertical direction. Here we expect the stability to be higher and mixing to be poor. The stability of the atmosphere as determined by the temperature gradient can change with the time of the day (sunlight and cloud cover) and has a strong influence on the dispersion, mixing and concentration levels in a region.

The smoke or gases emitted out of a chimney stack of a factory spread out in the form of a plume. This frequently shows an irregular form and shape which is difficult to predict. This prediction would require the inclusion of the time dependent nature of the flow, i.e. velocity (all three components which can be time dependent) and molecular diffusion effects into our mathematical description of the system. The flow is likely to be turbulent and so the velocities present would have different length and time scales. This means that there will be high frequencies and low frequencies present in the variable. It is numerically challenging to include all of these effects. The computational power required to determine the solution to the equations which describe this system is usually large. This arises when we use a pure and rigorous scientific approach to address the problem.

Molecular diffusion represents the spreading of a solute in a solution in the absence of bulk motion or convection. When a coloured dye is injected in a solvent it spreads slowly and the solution eventually becomes coloured with a uniform concentration. Take a cup of water to which you add a spoon of sugar. If you leave the cup as it is and wait for a long time the concentration of sugar becomes uniform in the solution. This spreading of sugar arises from *Brownian motion* in the solution. In the absence of velocity in the liquid this is a slow process and takes a long time. If we stir this solution it attains a uniform concentration much faster.

Stirring accelerates the diffusion or spreading. In a similar manner, the turbulence in the flow field of the atmosphere serves to mix the fluid. This turbulence is characterized by velocity fluctuations. The spread of the plume of the gases coming out of a chimney stack or coming out of the exhaust pipe of a vehicle is due to dispersion induced by the fluctuations of the turbulent velocity field and the concentration field. Thus, turbulence in the velocity, giving rise to the velocity fluctuations results in the shapes of the plumes of the gases emitted and its spread in the direction perpendicular to the flow. We can interpret this as acting in a manner similar to diffusion. This effect is called *dispersion* and is not a molecular property but depends on the flow field. It is much stronger than the molecular diffusion.

130 Introduction to Chemical Engineering

As engineers, we are interested primarily in ball park estimates of the quantities of interest. For instance, we may not need to know what the concentration levels are exactly but rather whether they exceed a threshold level or not. Hence, the high resolution level predicted by a detailed model may not be of any use. In addition to this information on the level of detail of the velocity fluctuations, etc. may not be available for incorporation in the model.

To get good estimates of the air quality in a region or the behaviour of such systems in general, engineers use a semi-empirical approach. It involves making simplifications and approximations which help simplify the rigorous scientific approach so that it can predict the behaviour of the real system accurately enough for the engineering application. This approach is usually used when we have systems whose interactions have not been fully understood. We illustrate this with an example of the spread of a plume from a chimney stack.

In this problem the concentration of the species depends on space and time. The spatiotemporal evolution of concentration is governed by convective and diffusive mass transport. Diffusion of mass is similar to heat conduction but now convection is also included. It is given by the following partial differential equation.

$$\frac{\partial c}{\partial t} + u\frac{\partial c}{\partial x} + v\frac{\partial c}{\partial y} + w\frac{\partial c}{\partial z} = D\left(\frac{\partial^2 c}{\partial x^2} + \frac{\partial^2 c}{\partial y^2} + \frac{\partial^2 c}{\partial z^2}\right) \qquad (6.1)$$

In this equation, chemical reactions in the atmosphere are neglected. Here u, v, w represent the velocity components in the x, y, z directions and c is the concentration of the pollutant. This equation is also based on the principle of conservation of mass. Here we have applied the mass balance to an infinitesemally small element (an infinitesemal control volume) and have obtained the various terms. The term with the time derivative represents the accumulation term, the terms with the velocity represent the net outflow term because of convection or wind velocity, and the second derivative term accounts for the diffusive transport of a species. In deriving the above we have neglected reactions taking place in the atmosphere.

In the statistical approach towards turbulence the instantaneous variables are written in terms of a time-averaged quantity and a fluctuating component.

$$u = \bar{u} + u', \; c = \bar{c} + c'$$

$$v = \bar{v} + v', \; w = \bar{w} + w'$$

This implies that the time average of the fluctuating variables is zero.

$$\frac{1}{\tau}\int_0^\tau u'(t)dt = 0 \qquad (6.2)$$

The wind velocity governing the convective transport is characterized by fluctuations in both magnitude and direction over a period of time. It is never a constant and cannot be controlled. The spreading of the plume that we observe coming out of the chimney stack arises from to these fast fluctuations. The concentration levels that are measured experimentally in a region represent a time-averaged value. This average is taken over a time interval which is in between the time scale of the turbulent fluctuations and the entire time scale of observation. If the time interval over which the average is taken is very large, then the average measured will smear out any macroscopic drifts or patterns present. If the time interval is too small, then the measured values will reflect the microscopic fluctuations.

To illustrate this, let us consider a slowly varying periodic curve on which we have superposed fast fluctuations. By slow variation we mean the time period of the oscillation is very large. When averaging is done over the entire time scale of measurement it gives one value, the average over the entire length. When it is done over an interval which is larger than the scale of fast fluctuations but lower than the time scale of the slower fluctuations we see a smoothly varying periodic function. What we have done is essentially filtered out the fluctuations or the noise. This situation arises in signal processing where a dominant signal is present and is corrupted by noise. Here Fourier transform techniques are used to identify frequencies present in the signal and these are used to remove the noise.

We explain this graphically. Figure 6.1 shows an actual variable being measured which has fluctuations superposed on the true value. It can be seen that the actual signal is periodic but noisy. In Figure 6.1(b), the signal is time averaged over the entire length of the measurement. As a result the average is only a single value. The periodic variation is lost. In Figure 6.1(c), the signal is time averaged over an interval which is greater than the time scale of the fluctuations but smaller than the time scale of the full measurement. This helps retain the macroscopic periodic variation.

Substituting in Eq. (6.1) for each variable in terms of the average and fluctuating component, we obtain

$$\frac{\partial(\bar{c}+c')}{\partial t} + (\bar{u}+u')\frac{\partial(\bar{c}+c')}{\partial x} + (\bar{v}+v')\frac{\partial(\bar{c}+c')}{\partial y} + (\bar{w}+w')\frac{\partial(\bar{c}+c')}{\partial z}$$

$$= D\left(\frac{\partial^2}{\partial x^2} + \frac{\partial^2}{\partial y^2} + \frac{\partial^2}{\partial z^2}\right)(\bar{c}+c') \tag{6.3}$$

Taking the time average of this equation and using Eq. (6.2), we obtain

$$\frac{\partial \bar{c}}{\partial t} + \bar{u}\frac{\partial \bar{c}}{\partial x} + \bar{v}\frac{\partial \bar{c}}{\partial y} + \bar{w}\frac{\partial \bar{c}}{\partial z} + \overline{u'\frac{\partial c'}{\partial x}} + \overline{v'\frac{\partial c'}{\partial y}} + \overline{w'\frac{\partial c'}{\partial z}} = D\nabla^2 \bar{c} \tag{6.4}$$

Equation (6.4) represents how the time-averaged concentration changes with space and time. This equation has the time average of the product of fluctuations in velocity and concentration derivatives as the third term on the left-hand side.

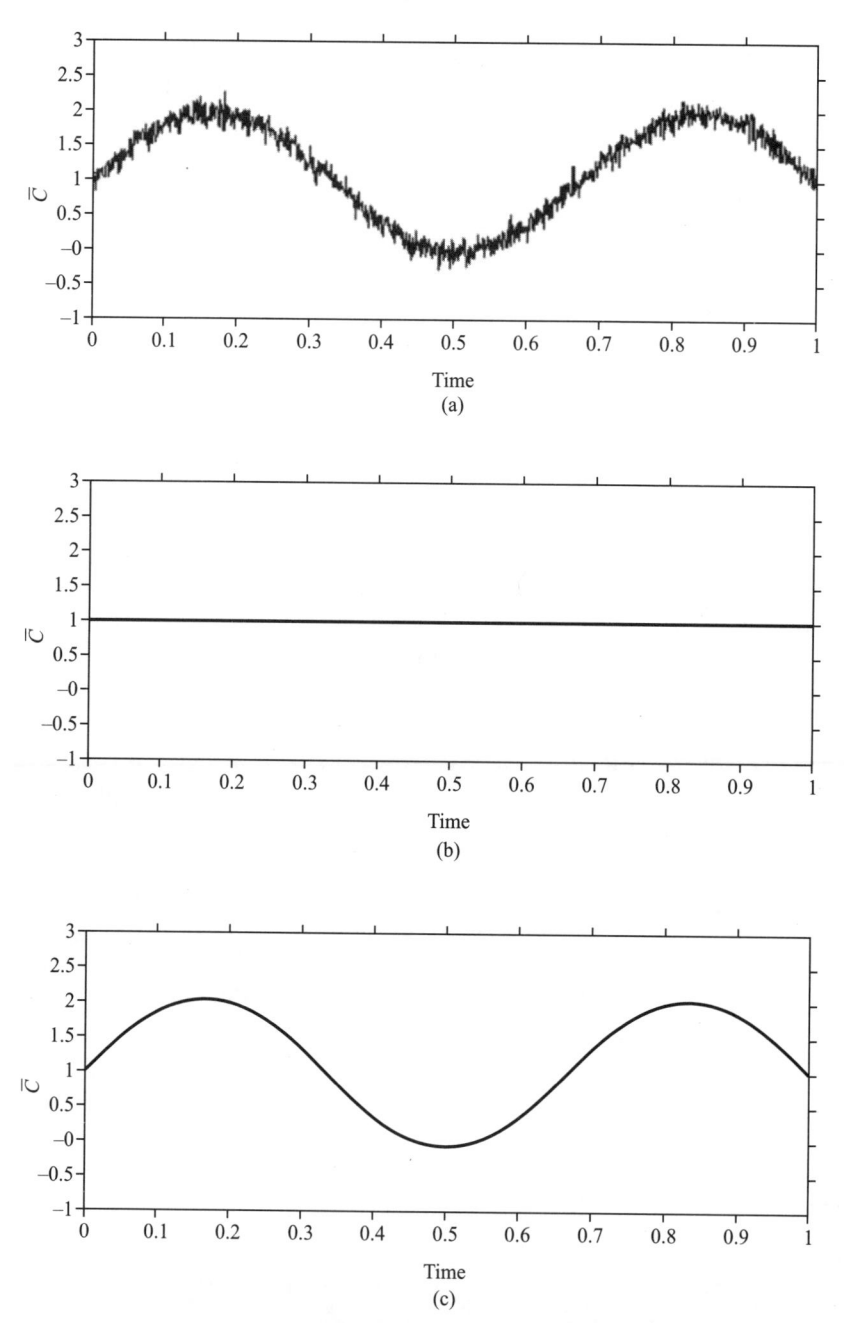

Figure 6.1 (a) Example of a time-varying quantity with fluctuations, (b) time-averaged signal when the interval of averaging is taken over the entire duration of the measurement, and (c) time average-signal when the time interval for averaging τ is larger than the time scale of fluctuations but lower than the measurement duration.

This term cannot be measured as usually the fluctuations are not measured and only the time-averaged values are measured. Besides in Eq. (6.4), an equation which governs the average concentration and velocities has a term with fluctuations of these quantities and so it cannot be solved. We would need independent equations for fluctuating variables. These are not available and this is called the *closure problem*. Let us use how this is overcome by engineers.

In the time-averaged equation the term containing the fluctuations of the concentrations has to be evaluated. These fluctuations arise from the turbulent velocity fluctuations. Physically, we anticipate that the fluctuations will give rise to spreading of the concentration as seen when dispersion is present. Hence, the term with the fluctuations represents the dispersion of gases which arises from the correlation between the fluctuations of the velocity and the fluctuations of the concentration. This term has an effect similar to molecular diffusion and can, hence, be represented in terms of a dispersion coefficient. The dispersion coefficient is a property of the flow field as opposed to the diffusion coefficient which is a molecular property.

As explained physically before, these are responsible for spreading the species by acting in a manner similar to diffusion. As this depends on the flow field, it is called the *dispersion* and the term containing the fluctuations is approximated as

$$\overline{u' \cdot \nabla c'} = -D_{\text{disp}} \nabla^2 \overline{c} \qquad (6.5)$$

Using this approach, we represent the terms containing the fluctuations due to concentration and velocity approximately in terms of the average concentration.

This form represents the transport due to fluctuations is similar to molecular diffusion. However, this is not a molecular property. Here the D_{disp} represents an effective dispersion term and is representative of all the effects which we have not considered, for example, stability of the atmosphere, presence of buildings, etc. Thus, this D_{disp} is a function of all the above variables. So far [up to Eq. (6.4)] the exercise of modelling has been mathematically rigorous. When we make the departure of approximating the terms containing the fluctuations by an effective dispersion we deviate from scientific rigour.

The correlations between the velocity field and the concentration field take into account the effect of spreading in the direction perpendicular to the flow or bulk velocity. The representation of this in terms of the dispersion coefficient is an approximation where we have used our physical insight to quantify the unmeasured fluctuations. However, we now have another problem, how do we measure this D_{disp} the dispersion coefficient.

Experiments have to be carried out to determine how this dispersion coefficient varies with the flow field and other conditions. Once this is obtained we use this value of the coefficient in the model and make predictions. This approach helps us get a mathematically closed system; an equation containing only time-averaged concentration, an equation whose solution can be found. The solution determines the dependence of the average concentration on space and time.

This approach is termed *semi-empirical* as it uses both fundamental sciences and empiricism (since the equations are based on conservation of mass and we have done time averaging of the equations). The problem of fluctuations being not measured is overcome from our physical insight where we approximate it in terms of a dispersion term. This term is obtained from realistic experiments or field data. The fact that D_{disp} is estimated from experimental data helps us come up with reliable predictions using the approximate model. The combination of science and experimental information helps instil confidence in our approach.

A fully empirical approach is one which is based purely on experiments. The information generated is valid for a particular set-up and set of operating conditions. But in this approach the ability to generalize and predict the behaviour under a different set of conditions is lost. In the fully empirical approach only interpolation can be performed confidently while semi-empirical approach also allows extrapolation with confidence.

A fully scientific approach may not be feasible since the level of understanding of various phenomena that are involved is not sufficient. Besides, as mentioned, the mathematical complexity may be very high and unnecessary. Hence engineers use a mix-and-match approach.

The above application shows how a semi-empirical approach can be effective in analysing engineering problems where a thorough understanding of all phenomena does not exist. A similar situation arises in several areas of momentum transfer, heat transfer, and mass transfer.

Applications in Turbulent Flows

As seen in Chapter 5, the pressure drop in a pipe where the flow is laminar can be determined analytically as being dependent on Re. In this case the exact functional dependency can be obtained from the momentum balance equation. When the flow is turbulent we know it is dependent on Re again but we do not know the functional form of the dependency. For this we need to carry out experiments and determine the exact functional relationship between the pressure drop and Re. Thus, science tells us that the pressure drop depends only on Re but the exact dependence has to be obtained from experiments. The scientific basis gives us the confidence that the friction factor does not depend on any other parameter. This again shows the integration of rigorous science and experiments to get a complete understanding of system behaviour.

As another application consider the modelling of turbulent flows. While the area of turbulence is interesting and challenging and of significant engineering importance, it is difficult to predict the behaviour of a turbulent flow accurately. The evolution of the time velocity at a point cannot be predicted by rigorous momentum balance in turbulent flows since all the information about the flow is not available. By this we mean the fluctuating velocity components, etc. are difficult to predict. However, in many problems the interest is in time-averaged values. We follow the same approach as in the atmospheric dispersion problem. Flow variables are written as a mean value over which a fluctuation is superposed as before. Our focus is on describing the mean behaviour in turbulent flow. We time-average the equations and see again that the correlation between the velocity fluctuation terms arise. It can be shown that the momentum balance in fluid flow is governed by

$$\frac{\partial u}{\partial t} + u \cdot \nabla u = -\nabla p + \mu \nabla^2 u \tag{6.6}$$

Writing $u = \bar{u} + u'$, $p = \bar{p} + p'$ and time-averaging, we obtain

$$\frac{\partial \bar{u}}{\partial t} + \bar{u} \cdot \nabla \bar{u} + \overline{u' \cdot \nabla u'} = -\nabla \bar{p} + \mu \nabla^2 \bar{u} \tag{6.7}$$

The third term on the left is written as

$$\overline{u' \cdot \nabla u'} = -\mu^{(t)} \nabla^2 \bar{u} \tag{6.8}$$

where $\mu^{(t)}$ represents the eddy or turbulent viscosity. This is not a molecular property but depends on the nature of the flow. The time average of the product of the fluctuating components can be interpreted to be in the form of a shear stress. An additional stress is generated in turbulent flows which acts besides the shear stress arising from fluid viscosity for laminar flows. Consequently, turbulent flows are modelled as liquids having the molecular viscosity and the eddy viscosity. The area of turbulence relies on using different approaches for estimating this eddy viscosity.

Now we know that engineers use a semi-rigorous approach to solve complex problems arising in several areas. The complexity arises from lack of information about the processes as well as from the computational difficulty in solving these problems. The semi-empirical approach is based on rigorous science. Approximations are made using physical intuition. In this process additional parameters are generated like D_{disp}, $\mu^{(t)}$ which are obtained from experiments. The two examples of dispersion and turbulence have been chosen so that the students can relate to the systems. These examples also show that there is a similarity in the transport of mass and momentum in turbulent flows. In solving engineering problems the two-pronged approach of using mathematical modelling and experiments is necessary. This way the drawbacks of one can be overcome by the other.

Exercises

1. The turbulent viscosity is not a property of fluid. This depends on flow conditions and the geometry. Different theories exist for estimating this. Two classical theories are the Prandtl mixing length theory and the k-epsilon model. Read how they are used to estimate the turbulent viscosity.

2. Dispersion coefficient is used in atmospheric pollution modelling to take into account the uncertainties induced by turbulence. Find out the parameters on which this depends and how this is estimated. What are the typical assumptions made?

   (**Hint:** See Pasquill stability criterion.)

7

Safety, Health, Environment and Ethics

Introduction

When the acronym SHE was written on the board and the students were asked what it stood for: a common answer from the students was *standard hydrogen electrode*. In the context of chemical process industry, SHE stands for *safety*, *health* and *environment*. These three words are interlinked and are extremely important in the corporate world. The industry which has a culture of a strong programme in SHE, will be able to retain its employees as opposed to another with a poor SHE programme. A vibrant and enthusiastic SHE programme sends out a strong message about the commitment of the industry towards encouraging safe practices and being environmentally friendly to the society as a whole and the employees in particular. It will have a good standing in society.

Safety in Chemical Process Industries

Chemical process industries are vulnerable to accidents. This does not imply that they are high-risk industries. A proper understanding of how the various systems behave and the processes occurring in these systems is important to ensure safe operation. Most accidents arise from human error or a lack of training of personnel and a poor understanding of the processes in each unit and the interaction between the different units. It can also be due to an improper or incorrect design of equipment or poor safety controls in different equipment. Many a time when a repair or maintenance work is undertaken modifications are carried out and these may compromise on the safety of the operation. An accident in the process can result from a safety procedure being overlooked and violated. This can lead to a release of toxic chemicals in the environment and this in turn has an adverse health impact on the population living in the neighbourhood of the plant. After the release these chemicals can persist in the atmosphere and have an adverse impact on health. It is, hence, necessary to study these three aspects simultaneously as they are interlinked.

> **Note to the faculty:** Several major accidents have occurred in the chemical industry in the recent past. Information on these is available on the Internet. Lessons have to be learnt from each of these accidents. One option is that the faculty should discuss these accidents with the student. The interest of students can be retained for a couple of cases at best this way. A faculty with an industrial experience may be able to do slightly better. Another option to make students aware of the inherent hazards is to divide the class into groups of 3 or 4 students and have them collect information on an accident. This will help them analyse an accident, the causes, the effects and the lessons learnt. The members of the group would learn to work together as a team.

Each group of 3 or 4 students can be asked to submit a report and make PowerPoint Presentations (PPTs). They would find out facts about specific accidents which have taken place in the past and present the details both in writing and orally. A general outline of points to be addressed in the report and presentation may include:

1. a discussion of the normal plant operation with specific reference to the area or units where the accident occurred;
2. the events which led to the accident (were there any modifications done to the plant);
3. identifying the cause of the accident: was it due to a fault in the design? Was it due to human error? Was it due to a poor safety culture? Was it caused by absence of administrative structure so that decisions could not be taken?
4. the impact of the accident, loss of lives and financial damage sustained.
5. the lessons learnt from the accident: what was the impact of the accident? Did it give rise to changes in plant operations? Did it change the attitude of the management?
6. identifying a similar plant with a good track record on safety and check if suitable safety measures have been implemented.

The presentations may be for 10 minutes. Each student in a group is given two minutes for making his part of the presentation. The students learn how to prepare slides and present material in an orderly manner. The order of the presentations by the member teams is randomly decided by the instructor. This will ensure that every member is prepared to present any aspect of the presentation and that all members work together in making the presentation.

The accidents may include the following:

- Flixborough disaster (1974)
- Disaster in Beek propylene plant (1975)
- Seveso disaster Italy (1976)
- San Juan Ixhuatepec disaster, Mexico City, (1984)
- Bhopal disaster (1984)
- Basel Chemical Spill (1986)
- North Sea Occidental (Piper alpha) (1990)
- Ufa train disaster, Russia (1989)
- HPCL refinery Vishakapatnam (1997)
- El Paso pipe line explosion Carlsbad, New Mexico (2000)
- University of California, Los Angeles, Accident in Chemistry Professor Patrick Marran's Lab (2008)
- Chernobyl nuclear disaster (1986)
- 3 Mile Island disaster (1979)
- Fukushima disaster (2011)
- BP Deepwater horizon oil spill (2010)
- Ships collision of Bombay coast (2010)
- Indian Oil Corporation, fire in oil depot, Jaipur (2010)

The presentations will help increase the awareness of students by making them think through the accident. They could identify the reasons that caused the accident and learn that the causes spanned a wide spectrum. They could learn how the violation of certain precautionary steps and a lack of understanding of the entire process are important factors for causing accidents. The accidents that have been reported are the ones which have resulted in significant damage to lives, property, and environment.

Everyday in different factories there are situations where a fault or a problem arises; for instance, a pump may not work, there may be a leak in the pipe, a boiler may not work, a pipeline could be choked preventing liquid from flowing. There could be a fire in a small area. Engineers and plant personnel are trained in handling these situations and ensure that the problem is addressed and taken care of suitably. They would possibly confine the problem to a small region and ensure it does not spread to other parts of the plant. These comprise a significant proportion of the situations, i.e. 99.99%. It is the remaining situations which comprise a very small percentage which draw the attention of the people as their adverse impact is high. The rate of accidents has come down because of the improved knowledge of the processes taking place in the systems, improvements in measurement techniques and control of disturbances, stronger legislations which ensure protocols are followed for the safety of personnel.

A few general lessons learnt from these disasters are:

(i) Ask if alternate routes are available for the manufacture of chemicals which are safer and more environmentally friendly.
(ii) See if temperature and pressures in the various vessels can be lowered thereby making them safer.
(iii) Reduce the inventory of chemicals so that at any given time the amount of the hazardous chemical is small and, hence, the likely damage that can be caused is minimal.
(iv) Periodically check the alarm system and other safety devices which are installed to control the behaviour of processes as part of regular maintenance.
(v) Ensure the standard of engineering is high and that the personnel available are well qualified and well trained.
(vi) Modifications should be allowed only after assessing their potential effects. The modifications can be in equipment design, flow sheet or even operating conditions.
(vii) The layout of the plant and its design are important. For instance, the control room should be away from potential high-risk areas.
(viii) Do not modify standard procedures especially over weekends or at the end of the day when you may be in a hurry.
(ix) Ensure plants are located away from residential, well-populated areas.
(x) Avoid situations where all protective mechanisms are simultaneously out of service.
(xi) Ensure there is a redundancy in protective systems and make sure that they are inspected and maintained so that any faults can be repaired.
(xii) Carry out analysis of consequences and possible extent of damage. Design and operate the plant to minimize these effects.

Lessons for the Management

(i) The management should show its sincerity in observing safety procedures to gain confidence of its employees. It should not act only when there is a threat to the plant to be shut down or if it affects the profit.
(ii) After an incident there should not be a witch hunt or search for a scapegoat amongst the junior-level staff.
(iii) Identify the real source of the problem.
(iv) Go with an open mind to understand the problem and only then solutions will emerge.

(v) Involve all stakeholders in the actions. These would be government officials, employees, neighbourhood residents, union leaders. Let them express their points of view and see how best their concerns can be addressed.
(vi) In a democracy literacy rate is high and everyone is educated. The start of new projects, expansions of existing projects, etc. have implications on the environment. There could be a technical impact on the environment or it could even be a socio-economic impact in the neighbourhood. The management has to ensure in these ventures that the concerns of all the parties are addressed. This includes the neighbourhood communities, the trade-union leaders, the employees at the higher levels, the officers and managers, politicians, the environmental regulatory agencies, and other legislative bodies.
(vii) Make sure that the organizational structure is such that there is an authority to take decisions. This will prevent the problem from escalating for want of a person authorized to taking a decision.
(viii) Have an external independent agency review and inspect steps taken to ensure safety of plants.

In spite of best efforts where these stakeholders are taken into confidence, there could be vested interests which may hinder the plans of the management for sustainable growth from being carried out. Typical examples of this include the Posco Steel plant in Orissa, the Nano car project in Singur (West Bengal) and Trivandrum Titanium Pvt. Ltd. In the first two, the socio-economic impact on the displaced farmers and in the third, the effluent treatment practices were issues which raised questions and disturbed the plant operation.

Importance of Quantitative Information

A scientist reported in the literature the operating conditions of conducting a reaction. He had suggested that the two reactants be mixed with a gentle warming to carry out the reaction smoothly, in his article. When a scientist from another country wanted to verify this he carried out the reaction following the same protocol. However, he found that the reaction was very violent and there was a small explosion in his laboratory. Luckily there were no casualties. Can you guess what went wrong? How is it that the same experiment and reaction behaved in two different ways?

The scientist who reported the experiment was from a cold Scandinavian country. The ambient temperature there was very low close to zero or sub-zero. Hence, in order for the reaction to occur it was necessary that the contents be slightly warmed. As you know the reaction rate dependency on temperature is given by the *Arrhenius relationship*. This dependency on temperature is exponential. The gentle heating was necessary in the Scandinavian country to raise the temperature. This ensured that the reaction could occur at a reasonable rate. The rate of heat loss to the surroundings was high since the ambience was cold. This ensured that the temperature did not rise in the reactor to unprecedented high levels. However, in a tropical country like India the room temperature is normally much higher, around 25–40°C. The reactants on mixing are also at this temperature and, hence, the reaction between the two can proceed at a reasonable rate without any gentle heating. Besides, the heat transferred or lost to the environment would be much lower as the ambient temperature is higher. The temperature difference between the reactant and the environment is reduced and this would lead to a rapid accumulation of heat and a drastic rise in temperature.

How could this problem have been avoided? The problem arose since all the information was not specified in the report. In particular, quantitative information was missing regarding the initial temperature and the ambient temperature. It is, hence, necessary to specify quantitative information so that protocols can be followed clearly without any ambiguity.

Luckily, there were no casualties in this incident. What favoured this is the low inventory in the reactor, so that there was not sufficient mass in the reactor to cause significant damage.

Can you guess if this reaction was endothermic or exothermic? It may appear that it is endothermic since heat was supplied in the experiment, however the drastic increase in the temperature indicates that the reaction is exothermic. The amount of heat supplied initially was to raise the temperature to ensure the reaction proceeded at a sufficiently high rate. If this heat is not supplied, then the reaction would occur incredibly slowly and we would have to wait for a long time to obtain our products. The fact that the reaction blew up when carried in a tropical country again indicates that the reaction is exothermic as it confirms that it was accompanied by a significant amount of heat release. The nature of the reaction being exothermic or endothermic is decided by thermodynamics. How fast the heat is released is decided by the rate of the reaction.

Le Chatelier's principle states that if the reaction is exothermic, the equilibrium shifts to the left when heat is supplied to the system. However, in the above case we are considering an irreversible reaction. Here we are not talking about the equilibrium state attained but about the rate of the reaction. The rate of the reaction and the exothermicity decide the rate of temperature rise as the reaction progresses. Nuclear reactions are also exothermic and can also exhibit runaway behaviour. This was the cause of the accidents in Chernobyl and Fukushima.

We now discuss different case studies to illustrate different aspects of chemicals which can adversely affect ecology and environment.

Case Study 1: Extinction of Different Species of Vultures

India has been experiencing a drastic decline in the population of vultures since the last decade. It is estimated that over 100,000 birds have become extinct from several of their strongholds in urban and rural areas as well as protected areas and forests. Of the eight species of vultures found in the subcontinent, the white-backed vulture (*Pseudogyps bengalensis*) and the griffon (*Gyps fulvus*) Indian (*G. indicus*) and Slender-billed (*G. tenuirostris*) have been severely affected. Several other species have also suffered a proportionate decline in population. The vulture plays an important role in the disposal of carcasses, which are often left to rot, and are important for reducing the risk of diseases. The decline in vulture population started around 1990, became visible by 1995 and turned serious after 1997.

In the Delhi–Agra–Bharatpur triangle, and neighbouring localities, vulture populations slid from approximately 20,000 birds in 1990 to about 150 in 1999. The cause for this decline was not known then and scientists were puzzled. It was only around 2004 that the cause was identified as the drug, diclofenac, an anti-inflammatory agent, used in veterinary medicine as a painkiller in specific livestock. Cows treated with this drug could overcome their pain and work for longer hours in the fields. Vultures ingested this drug when they fed on carcasses of animals treated with this drug. The drug acts as a fatal poison to vultures and they die from kidney failure. Surveys in India showed that the country's Indian and slender-billed vulture populations declined by almost 97 per cent between 1992 and 2007. White-rumped vultures fared even worse, dropping by 99.9 per cent, to just one thousandth of their 1992 population.

The drug is now proven to have been responsible for the near-total extinction of three species of vulture in south Asia—white-rumped *Gyps bengalensis*, Indian *G. indicus* and Slender-billed *G. tenuirostris*. These species formerly among the commonest large birds of prey in the world are now locally extinct in several regions.

Diclofenac is widely used in human medicine globally, but was introduced to the veterinary market on the Indian subcontinent during the early 1990s. The drug is inexpensive and widely used in the treatment of inflammation, pain and fever in livestock.

Vultures provide a crucial service in the ecosystem through the disposal of carcasses. The decline in their population has had a significant socio-economic impact across the Indian subcontinent. Without vultures, hundreds of thousands of carcasses have gone uneaten—left to rot in the sun. These pose a serious risk to human health. Carcasses provide a breeding ground for numerous infectious diseases and encourage the proliferation of pest species such as rats. The absence of vultures has resulted in an explosion in the number of feral dogs—the bites of which are the most common cause of human rabies. A recent study in India estimates that, concurrent with the vulture die-off, there has been an increase in the feral dog population of at least 5.5 million. It is calculated that this has resulted in over 38.5 million additional dog bites and close to 47,300 extra deaths from rabies. It is estimated that the increase in the number of rabies' victims may have cost the Indian economy $34 billion.

In 2006, the Governments of India, Pakistan, and Nepal finally introduced a ban on the manufacture of diclofenac for veterinary applications, and pharmaceutical firms are now encouraged to promote an alternative drug, meloxicam, which is proven to be safe for vultures.

The decline of Asian vultures is one of the steepest declines experienced by any bird species. A 30 per cent decrease in the vulture population can occur when even one in 760 carcasses contain diclofenac at a dose lethal to vultures.

Main Reasons for Vulture Deaths

Diclofenac in livestock

Diclofenac was used to treat cattle as it was an inexpensive painkiller. Scavengers feed on dead cattle administered with diclofenac. The number of endangered South Asian vultures being killed by the painkiller has declined substantially after its ban on veterinary use. This has raised hopes that the species in the region can be saved from extinction. The initial death rate from diclofenac was high. It is felt that more significant steps need to be adopted to eradicate the illegal use of the drug. This would prevent the number of birds being poisoned from creeping back up. Figure 7.1 shows the structure of diclofenac.

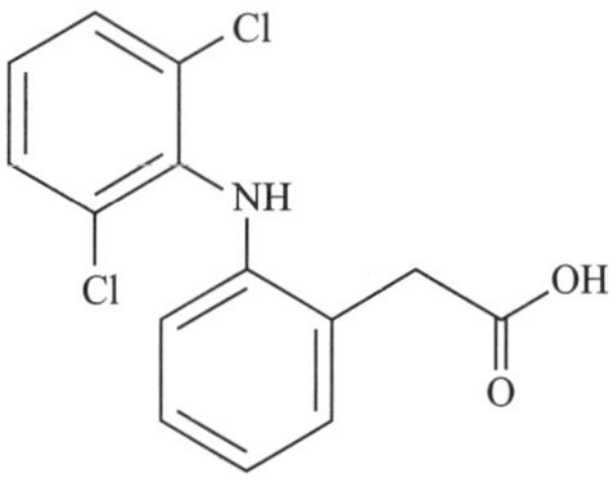

Figure 7.1 2-(2-(2,6-dichlorophenylamino) phenyl) acetic (structure of diclofenac).

Pesticides

Many farmers spray cattle carcasses with pesticides such as organo-chlorine and Organo-phosphorous compounds to prevent them from spreading foul odour. This pesticide infested carcass may be eaten by vultures leading to their death. Ecologists have found instances where hundreds of vultures have died this way.

This example shows how the ecosystem is an intricate network. The balance of the ecosystem has to be preserved. This can be upset by actions which are taken and which do not have a seemingly direct effect on the ecosystem. Specifically, the introduction of diclofenac as a painkiller helped alleviate the pain of the cattle but caused other problems: decline in vulture population and increase in rabies' victims.

Steps That Need To Be Taken

A complete ban on the usage of diclofenac as a veterinary painkiller has to be implemented effectively and awareness about alternative harmless drugs like meloxicam has to be spread among cattle owners.

Breeding sites have to be monitored to check if any dead vulture is found in the neighbourhood to check the efficacy of the ban. Discussions with cattle-keeping communities in feeding areas must be conducted. They may have to be paid not to sell the old cattle to slaughterhouses. They should let old cattle die without intake of harmful drugs like diclofenac. These steps can also help prevent vulture deaths.

Case Study 2: DDT

DDT 1, 1, 1-trichloro-2, 2-bis (*p*-chlorophenyl) ethane (structural formula in Figure 7.2) used to be a well-known insecticide in the 1960s. It was particularly effective against mosquitoes.

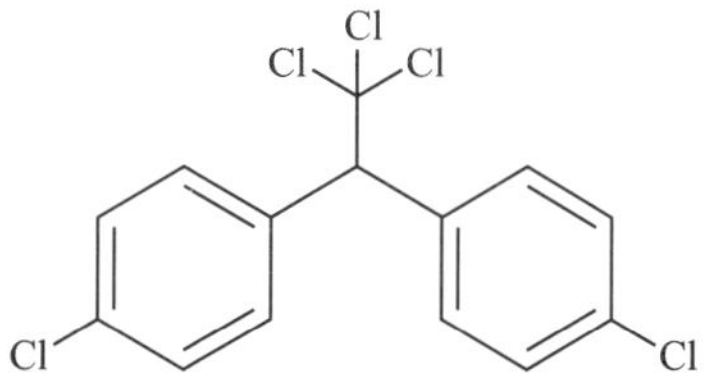

Figure 7.2 The structual formula of DDT.

DDT is a synthetic white, powdery, waxy hydrophobic substance. It does not dissolve in water and so does not contaminate it, but readily dissolves in solvents and oils. It is applied as a white smoky mist. DDT's hydrophobic nature is both a blessing and a curse. On the one hand, it does not contaminate water sources, which is good; while, on the other hand, it does not get dissolved and diluted into virtual nothingness. As a result, it remains in the environment for a really long time.

DDT is strongly absorbed by soil. Depending on conditions, its soil half-life can range from 20 days to 30 years. Its breakdown products, DDE and DDD, are also highly persistent and have similar chemical and physical properties. The formation of these products is shown in Figure 7.3.

The Early Years

The insecticidal properties of DDT were discovered by Swiss scientist Paul Müller in 1942. For this, he was honoured with the Nobel Prize. DDT is a mixture of isomers, principally p, p'-DDT, with lesser amounts of o, p'-DDT.

Its usage increased during the Second World War. It has been used to control the mosquito-borne malaria, and was used extensively as a general agricultural insecticide.

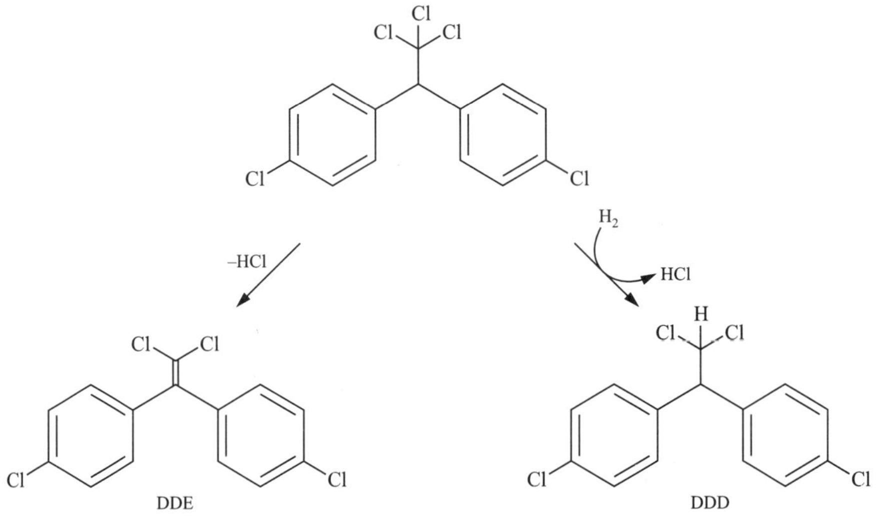

Figure 7.3 The formation of DDD and DDE from DDT.

How Does DDT Work as an Insecticide?

DDT kills insects by chemically enhancing the electrical connections between their neurons, short-circuiting them into spasms and death. In insects it opens sodium ion channels in neurons, causing them to fire spontaneously, which leads to spasms and eventual death. Insects with certain mutations in their sodium channel gene are resistant to DDT and other similar insecticides.

In humans, however, it is suspected to affect health through genotoxicity or endocrine disruption. The former implies that it affects the genes and is a potential carcinogen. The latter imples that it can affect the functioning of the endocrinal glands.

Advantages of Using DDT

DDT spraying helps control the incidence of malaria. The evidence from more than 50 years of use indicate that, if properly applied, it is not harmful to humans or the environment in general. Further, the evidence is that, when it is **properly applied**, mosquitoes do not become resistant to DDT, a benefit that is not shared by most alternative sprays.

DDT appeared to be very harmful in the 1950s and 1960s because of its widespread use in heavy dosages, mostly from government spray campaigns but also from overuse by private sprayers who did not pay attention to proper conservation principles. When DDT is sprayed in massive doses, birds can suffer acute and long-term effects. This can be traced to the massive agricultural use of DDT.

Initially, DDT was very successful in the control of malaria, as well as against agricultural pests. But by the 1950s, resistance problems had developed, and during the 1960s, a number of serious environmental problems were identified leading to wide-ranging restrictions on its use. Many insect species developed resistance to DDT. The first cases of resistant flies were known to scientists as early as 1947, although this was not widely reported at the time. In the intervening years, resistance problems increased mostly because of overuse in agriculture. By 1984, a world survey showed that 233 species, mostly insects, were resistant to DDT.

Due to its lipophilic properties, DDT has a high potential to *bio-accumulate*, especially in predatory birds. DDT, DDE, and DDD magnify through the food chain, with apex predators such as raptor birds concentrating more chemicals than the other animals in the same environment. They are stored mainly in body fat. DDT and DDE are very resistant to metabolism; in humans, their half lives vary from 6 to 10 years. This high half life make it fall in the category of persistent organic pollutant (POP).

Health Effects

DDT is moderately to slightly toxic to mammals. Of particular concern is its potential to mimic hormones and thereby disrupt endocrine systems in wildlife and possibly humans.

DDT is categorized by the World Health Organization as Class II "moderately hazardous".

DDT mainly affects the central and peripheral nervous systems and the liver. Acute effects in humans exposed to low to moderate levels may include nausea, diarrhoea, increased liver enzyme activity, irritation to the eyes, nose and/or throat. At higher doses, tremors and convulsions are possible. Deaths from exposure to DDT are rare.

It has been noted that DDT and its breakdown products are probable human carcinogens. DDT levels in breast milk in regions where DDT is used against malaria greatly exceeds the allowable standards for breast-feeding infants. These levels are associated with neurological abnormalities in babies.

Residues in Food

DDT is very fat soluble and is, therefore, found in fatty foods such as meat and dairy products. Eggs analysed as part of the Total Diet Survey in 1996 were found to contain residues as p, p'-DDT, o, p'-DDT, p, p-TDE and p, p'-DDE. The dietary intake of DDT is considerably high in developing countries.

Ecological Effects

The effect of DDT on the environment was first studied in detail by Rachel Carson in 1962. She authored *Silent Spring* and it contained a detailed analysis of the harmful effects of DDT on the ecology. This can be seen as the beginning of the environment movement. It brought out the fact that DDT could harm wildlife and could also have an adverse effect on human health. This started a strong lobby arguing for a ban on DDT. There was also another lobby in favour of the use of DDT to eradicate diseases. The former argued that the use of DDT resulted in an increase in the resistant species and this increased the number of victims succumbing to malaria. The latter argued that the use of DDT was important to control malaria and other diseases.

DDT and its breakdown products have widespread persistence in the environment, and a high potential to bio-accumulate. It is highly toxic to fish. Smaller fish are more susceptible than larger ones of the same species. An increase in temperature decreases the toxicity of DDT to fish. DDT and its metabolites can lower the reproductive rate of birds by causing eggshell thinning which leads to egg breakage, causing embryo deaths. Predatory birds are the most sensitive. Birds in remote locations can be affected by DDT contamination. In the United States, the bald eagle nearly became extinct because of environmental exposure to DDT. Other species which were affected were the peregrine falcon and California condors. There is strong evidence that p, p'-DDE inhibits calcium ATPase in the membrane of the shell gland and reduces the transport of calcium carbonate from blood into the eggshell gland. This results in a dose-dependent thickness reduction in eggshells. There is also evidence that o, p'-DDT disrupts female reproductive tract development, impairing eggshell quality later.

DDT, DDD, and DDE are all strongly suspected of being environmental endocrine disrupters. DDT can have reproductive endocrine effects and also has a major toxic effect on the adrenal glands. DDT-related deformities in birds include clubbed feet and crossed bills. There is also concern that it has the potential to disrupt the endocrine system of humans.

DDT Restrictions

Control actions to ban or severely restrict DDT have been taken by over 38 countries that began in the early 1970s. In at least 26 countries, DDT is completely banned, and in 12 others it is severely restricted. A total ban has been imposed in Canada, Chile, Korea, Poland, Singapore, and Switzerland. In these latter cases, it is permitted for use by government agencies for special programmes, usually involving vector control.

This case study points at the danger of using strong chemicals as insecticides. In the initial stages the effectiveness of the molecule may be very encouraging but the harmful effects manifest after a certain time has lapsed.

Case Study 3: Environmental Hazards of a Green Project

India is dependent heavily on fossil fuels to meet its energy requirements. Coal-based power plants give rise to atmospheric pollution. They release particulate emissions and gases such as carbon dioxide responsible for global warming. In a move to go towards renewable energy, the Government has encouraged research and is investing more in renewable sources such as wind energy and solar energy. In addition to being clean this reduces the dependency on fossil fuels. In this connection a 113 MW Andhra Lake Wind Power Project was approved and promoted by the Indo-German enterprise **Enercon India**. The project is located at a distance of 3.5 km from the Bhimashankar Wildlife Sanctuary. The project is spread over 14 villages of the Khed and Maval talukas in Maharashtra. It covers an area of 194.66 hectares of reserve forest land and is estimated to cost ₹ 772 crore. The company plans to erect 445 wind turbines in the Khed–Maval belt of Pune district.

Normally, no developmental activity is expected to take place within a 10 km radius of a national park. The permission given to Enercon India violates this. Enercon India Ltd. was given the permission to cut 26,615 trees but it is suspected that more than 3 lakh trees have already been cut in the region, to construct a 20 km long road along the mountains to reach the windmill site. The project started in March 2010. Till now, 40 of the 142 sanctioned windmills have come up across the region. Illegal tree cutting and blasting in the area as part of the project pose a danger to the wildlife there. Figure 7.4 shows the labourers engaged in blast activity. And Figure 7.5 shows a picture of a blasted and deforested hillside.

Figure 7.4 Labourers engaged in blasting activity for a 113 MW wind power project coming up in Maharashtra.

Figure 7.5 A picture of the blasted and deforested hillside.

The Adverse Effects of this Green Project

The damage caused by this green project to the environment is significant. Most trees felled are 30 to 50 years old, and include valuable species like mahua, beheda, jamun, chaar and wild mango, that villagers rely on as a source of extra income.

Due to the constant noise of blasting, several species of birds and the giant Indian squirrel have disappeared from the area. The wind turbines pose a great danger to the migratory patterns of birds.

The loose rubbles generated by the blasting has created a danger of landslides that can possibly destroy nearby farmlands, a scarce commodity in the hilly area. During the monsoons, the heavy rains in this area have the potential to carry all the rubble into the paddy plots.

Erosion from deforestation and blasting is likely to damage the catchment areas of the Bhima, Bhama, and Indrayani rivers and cause siltation in the reservoirs in the long run.

The project has now come under the scrutiny of the Western Ghat Expert Ecology Panel (WGEEP). The Bombay High Court has now restrained the promoter, Enercon (India) Limited, from cutting a single tree until further orders.

This case study shows the need to assess the impact of the overall project on the environment. The effect of deforestation as part of a green project based on renewable energy has a long-term adverse impact on the environment. The damage already caused may not be compensated by the greenness of the project. This project shows that development has to be sustainable in an overall sense. So, although we may end up getting renewable energy the damage to the environment may be severe. This case study empahsises the need for development to be sustainable.

Case Study 4: Endosulfan

Endosulfan is a chlorinated hydrocarbon insecticide and a caricide of the cyclodiene subgroup. It acts as a poison to a wide variety of insects and mites, including whiteflies, aphids, leafhoppers, colorado potato beetles and cabbage worms. It is used primarily on a wide variety of food crops including tea, coffee, fruits, and vegetables, as well as on rice, cereals, maize, sorghum, or other grains. Different formulations of endosulfan are possible and these include emulsifiable concentrate, wettable powder, ultra-low volume (ULV) liquid, and smoke tablets. Endosulfan is made up of a mixture of two molecular forms (isomers) of endosulfan, the alpha and beta isomers. Commercial names of this product include theodan, endocide, beosit, cyclodan, malix, thimul, and thifor. This is a chlorinated compound like DDT.

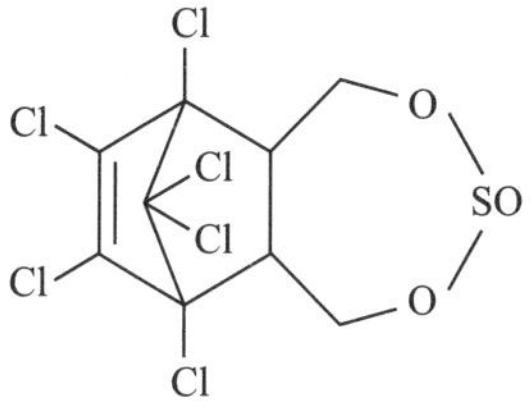

Figure 7.6 Structural formula of endosulfan.

Its molecular formula is $C_9H_6Cl_6O_3S$ and its CAS chemical name is 6,7,8,9,10,10-hexachloro-1,5,5a,6,9,-9a-hexahydro-6,9-methano-2,4,3-benzodioxathiepin-3-oxide.

This insecticide has been in the news in India since the beginning of the 21st century. It has proven to be a controversial insecticide and is similar in many respects to DDT.

The Governments of Kerala and Karnataka have been seeking a ban on the production and use of endosulfan. Kerala is one of the worst-hit states by the adverse effects of endosulfan. The details are now discussed.

Effects of Aerial Spraying of Endosulfan on Residents of Kasaragod, Kerala

From the mid-1970s, pesticide endosulfan had been aerially sprayed by the Plantation Corporation of Kerala on cashew nut plantation covering several villages like Muliyar, Periye, Enmakaje, Cheenmeni, and Padre in Kasaragod district, Kerala. The people residing in the villages within the plantation have been afflicted with different kinds of illnesses which, according to the villagers, were not present before aerial spraying was started on the cashew nut plantation. In these places many children who have been exposed to the aerial spray are called "living martyrs". For the last 20 years aerial spraying has been done in Enmakaje and the prevalence of unusual diseases is attributed to this.

152 *Introduction to Chemical Engineering*

The resident doctors have recorded that the symptoms resemble that arising from poisonous effects of pesticides. Congenital diseases in children, cancer, physical deformity, skin allergy are seen in other villages such as Padre, Periye, Cheenmeni, Bellur and Rajapuram. Several households in Padre village suffer from liver cancer, blood cancer, cerebral palsy, epilepsy, mental retardedness, asthma, and infertility. People suffered from suffocation during spray time and many inhabitants suffer from breathing disorders, irritation in the eyes, cough, headache after aerial spraying. In these regions unusual abortions among women and animals, birth of handicapped children, suicidal tendencies in people are prevalent. The death of fishes, honeybees, frogs, birds, chicken and even cows was found to coincide with the spraying. Figure 7.7 shows the effect of endosulfan on children and infants.

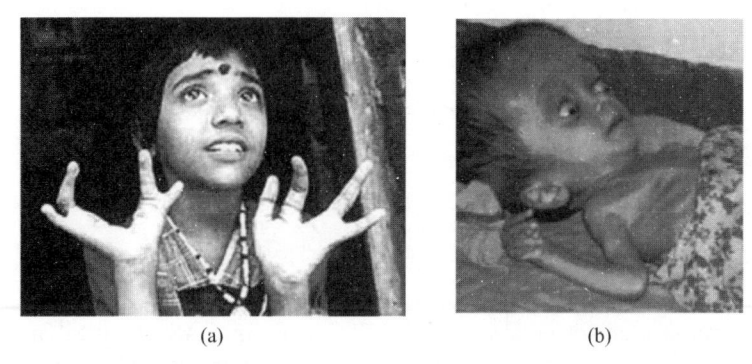

(a)   (b)

Figure 7.7 Effect of endosulfan on children and infants.

While about 500 deaths from 1995 have been officially acknowledged as arising from the spraying of endosulfan; unofficial estimates put the total number of deaths since the late seventies to around 4,000. The insect killer was sprayed aerially using helicopters. As the plantations are mostly in mountainous areas, the pesticide drains and gets washed down the slopes during the rains, contaminating the groundwater. It is suspected that consuming this water results in diseases ranging from physical deformities, cancers, birth disorders and damage to brain and nervous system. This can also cause mental retardation, cancer and infertility in the victims.

Effect of Endosulfan Use in Mango Orchards on Residents of Palakkad District, Kerala

Muthalamada panchayat in Kerala's Palakkad district has gained notoriety for excessive use of endosulfan in its mango orchards. Its effect is showing on the people living there. The pesticide is sprayed on individual trees with nozzles directed skywards. As a result of this, the person spraying gets affected directly.

Newborn children are infected with hydrocephalus in this region. Here there is accumulation of the cerebro spinal fluid in the brain. The cause of these health effects is similar to that of children in Kasaragod where the Plantation Corporation of Kerala aerially sprayed endosulfan on the cashew plantations for over two decades, leading to serious health problems. A survey of 9,000 households in Muthalamada and Kollengode panchayats in November, 2010 identified 46 suspected endosulfan victims. The victims were in the 0–14 age group. The emergence of health problems coincides with intense spraying of endosulfan that started 14 years ago.

Endosulfan was banned in Kerala in 2005 after the Centre issued a gazetted notification withholding the use of endosulfan in the state, on the basis of reports of the National Institute of Occupational Health and other committees. But that ban has been ineffective. Nearly 300 landholders of Palakkad who own big plantations in the region use endosulfan and other pesticides extensively during the flowering season to kill pests—leaf miners and leaf hoppers. Endosulfan is easily available across the district borders in Tamil Nadu where it is not banned.

India is one of the main producers and consumers of endosulfan in the world. In addition to Kerala, the states of Karnataka, Punjab, Assam, and Andhra Pradesh are afflicted with this problem.

Conflict between States and Central Government

The chief ministers of Kerala and Karnataka, two of the most affected states, have been insisting on a complete ban of endosulfan. However, the Union Ministry has turned down the request and has sought a scientific and epidemiological study on the side effects of endosulfan, to help the Centre decide on the ban. Endosulfan in India is manufactured by Hindustan Insecticides Limited, Excel Crop Care Limited, and Coromandel Fertilisers Limited. They produce around, 9000 tons of which around half is for domestic use and the remaining is for exports. The manufacture of endosulfan is an industry worth ₹ 300 crores (60 million US dollars).

Impact of Endosulfan

Environmental impact of endosulfan

The use of endosulfan generates endosulfan sulphate. This is released into water and gets absorbed in sediments and bio accumulates in aquatic organisms. Endosulfan may also persist in the atmosphere. Exposure to endosulfan can also arise from the ingestion of contaminated food as endosulfan residues have been found in numerous food products, vegetables, tobacco, milk, etc. Endosulfan may be found in surface water and air near the areas of application. Endosulfan residues are found on the ground too and they affect plant life. Endosulfan is highly toxic to fish species and aquatic invertebrates. It has been shown to severely affect liver histology in fish.

Impact on health of animals and human beings

Endosulfan is one of the most toxic pesticides in the market today, responsible for many fatal pesticide poisoning incidents around the world. Endosulfan is also a xenoestrogen—a synthetic substance that imitates or enhances the effect of estrogens—and it can act as an endocrine disruptor, causing reproductive and developmental damage in both animals and humans. Consumers can ingest endosulfan from residues in food.

Toxicological effects

(a) **Acute toxicity.** Endosulfan is a highly toxic substance and carries the signal word 'DANGER' on the label. Undiluted endosulfan is slowly and incompletely absorbed into the body whereas absorption is more rapid in the presence of alcohols, oils, and emulsifiers. Stimulation of the central nervous system is the major characteristic of endosulfan poisoning. Symptoms of acute exposure include a lack of coordination, even the loss of ability to stand. The other signs of poisoning include gagging, vomiting, diarrhoea, agitation, convulsions and loss of consciousness. Blindness has been documented in cows which grazed on a field sprayed with the compound. In an accidental exposure, sheep and pigs grazing on a sprayed field suffered a lack of muscle coordination and blindness.

(b) **Chronic toxicity.** Endosulfan has been found to cause high rates of mortality, liver enlargement, reduced growth and survival, changes in kidney structure, and changes in blood chemistry.

Reproductive and developmental effects

Endosulfan can also affect the human development. Researchers studying children from many villages in Kasargod district, Kerala, have linked endosulfan exposure to delays in sexual maturity among boys. Endosulfan was the only pesticide applied to cashew plantations in the villages for 20 years and had contaminated the village environment. The researchers compared the villagers to a control group of boys from a demographically similar village that lacked a history of endosulfan pollution. Relative to the control group, the exposed boys had high levels of endosulfan in their bodies, lower levels of testosterone, and a delay in reaching sexual maturity. Birth defects in the male reproductive system including cryptorchidism (absence of one or both testes) were also more prevalent in the study group. The researchers concluded that endosulfan exposure in male children may delay sexual maturity and interfere with sex hormone synthesis. Increased incidences of cryptorchidism have been observed in other studies of endosulfan-exposed populations.

A 2007 study by the California Department of Public Health found that women who lived near the farm fields sprayed with endosulfan and the related organochloride pesticide dicofol during the first eight weeks of pregnancy are several times more likely to give birth to children with autism. This is the first study to look for an association between endosulfan and autism. Additional studies are needed to confirm the connection. A 2009 study on epidemiology and rodents suggests male reproductive and autism effects are open to other interpretations, and that developmental or reproductive toxicity occurs only at endosulfan doses that cause neurotoxicity.

Mutagenic and carcinogenic effects

Endosulfan is not listed as a known, probable, or possible carcinogen by the (US) Environment Protection Agency (EPA), (IARC), or other agencies. There are no epidemiological studies linking exposure to endosulfan specifically to cancer in humans, but *in vitro* assays have shown that endosulfan can promote proliferation of human breast cancer cells. The evidence of carcinogenicity in animals is mixed.

The main conclusions are:

(a) The general population does not appear to be at risk from endosulfan residues in food. Exposure of the general population via air and drinking water is generally low.

(b) Occupational exposure has resulted in some incidents of poisoning. These appear to have occurred, however, only when adequate safety precautions were not taken.

(c) In terms of the general environment, endosulfan is highly toxic for some aquatic species, particularly fish. Endosulfan is moderately toxic for honeybees.

(d) Endosulfan does not accumulate in food chains and is eliminated rapidly from the body, therefore there is no danger of endosulfan poisoning to consumers of food crops using endosulfan as an insecticide. Moreover, the half life of endosulfan is a few days indicating that its effects will not be significant by the time the food crop reaches the consumers.

Endosulfan is being phased out globally. Endosulfan became a highly controversial agrichemical due to its acute toxicity and role as an endocrine disruptor. Because of its threats to human health and the environment, a global ban on the manufacture and use of endosulfan was negotiated under the Stockholm Convention in April, 2011. The ban will take effect in mid-2012, with certain uses exempted for 5 additional years. More than 80 countries, including the European Union, Australia and New Zealand, several west African nations, the United States, Brazil and Canada had already banned it or announced phase-outs by the time the Stockholm Convention ban was agreed upon. It is still used extensively in India, China, and a few other countries. It is produced by several manufacturers in India and China.

The United Nations Organization classifies endosulfan as a highly dangerous insect killer and it is banned in 62 countries. Despite this fact, for 26 years the Kerala Government sprayed the deadly pesticide in cashew plantations owned by the Plantation Corporation of Kerala. Presently, endosulfan has been banned from production and use for eight weeks in India since May 14, 2011 and a scientific study is being done to find out the effects of endosulfan in a detailed manner.

Most of the health problems have arisen from the excess spraying of endosulfan. The plantation wokers were also not informed of the precautions which need to be taken.

The High Court and the State Government have banned the use of endosulfan in Kerala. However, endosulfan happens to be illegally available and is used in Kasaragod, Kerala. Efforts to ensure that the ban of endosulfan is strictly implemented are being undertaken.

In the context of India, it is important to develop and implement community-based certification and local standards. The community must be made aware of the safety implications of the pesticide used.

The need for a comprehensive scientific epidemiological study will be able to establish the role of endosulfan on the health problems faced by the population. The question that arises is: should we wait for the results of this study? Is it worth risking and exposing the population to endosulfan until the study is concluded? It may be wise to be cautious and seek other alternatives as insecticides which may be more expensive but less harmful to human beings.

There is a strong similarity in the case studies of endosulfan and DDT. Both were welcomed as effective and inexpensive pesticides in the early stages. Later on it was found that there was a strong negative influence on the environment and the health of wildlife and humans. It also appears that this situation has arisen from the overzealous and rampant use of these insecticides.

Case Study 5: Plachimada Bottling Plant of Coca-Cola

Coca-Cola, the giant soft drink company, came to India in 1993 just when the economy of the country was being opened up. It was looking for a site to put up its plant. It needed a site where water was available in plenty. This was going to be a challenge when large areas in India are known to face an acute water crisis.

The liberalization of the economy and economic reform were going strong when Coca-Cola came in. The Central and state governments were all eager to attract these multinational companies in the hope that this would lead to the creation of employment, in their regions.

The effect of the advent of these companies could be seen on the urban front. These companies sponsored big sporting events. However, their positive effect was felt only in the top tier of society. Below this there were millions of Indians who were adversely affected. The Coca-Cola plant is an example of how below the layer of development and growth in the economy there was a huge damage to the environment.

Coca-Cola took over several Indian bottling companies, capturing their marketing and distribution systems. It appeared that India was on the verge of a big transformation. In the countryside, Coca-Cola's bottling plants were getting negative reviews; however, Coca-Cola had sound reasons for choosing Plachimada. It is in the heart of Kerala's water belt, with large underground water deposits. The site was between two large reservoirs and ten metres south of an irrigation canal. This site is surrounded by colonies of poor people who eked out their living as daily wage labourers.

The state of Kerala's "reform"-minded government ensured that the plant duly got a licence from the local council. Coca-Cola bought a property of 40 acres, built a plant, sank six bore wells, and commenced operations.

In a span of six months the local populace could see the water levels drop. The quality of water deteriorated giving people diarrhoea and skin diseases. Cooking of rice and lentils was rendered difficult. The quality of well water was affected over a radius of three to four kilometres.

The locals were mostly tribals. On April 22, 2002 the locals staged a peaceful agitation and the plant was shut down.

The old village wells were 150–200 feet deep. The company's bore wells went down to 750 to 1000 feet. As the water table dropped, the concentration of different toxic substances in the water increased. The plant was removing water from the groundwater table and in return was bringing in drums of sludge from the filtering processes of the plants. The company told the locals the sludge was good for agriculture and could serve as a fertilizer. This sludge made the people sick and people contracted skin diseases.

Lab analysis by the Kerala State Pollution Control Board showed dangerous levels of cadmium in the sludge. Another report done at Exeter University in the United Kingdom found in the water in a well near the plant not only impermissible amounts of cadmium but lead at levels that "could have devastating consequences", particularly for pregnant women.

The cruel truth was that water from the underground sources was being pumped out for free and sold back to the local people to make millions every day. This process resulted in the destruction of our environment and damage to the health of the local people. The local people viewed the multinationals as outsiders who have taken their water, filtered it and sold it back to them for a price.

In this era where sustainable development is a key mantra or slogan, growth has to occur in such a manner that there are no adverse implications on the environment. This simple fact was overlooked when allowing Coca-Cola to set up its operations.

Case Study 6: Marine Disasters in the Arabian Sea Near Mumbai

In August 2011, the bulk carrier MV Rak sunk in the Arabian sea at a distance of 20 nautical miles from the Mumbai port. The ship was carrying 60,000 tons of coal, 290 tons of fuel oil and 50 tons of diesel. The Indian Navy and Coast guard rescued all the sailors from the ship and, as a result, there was no loss of life. However, the fuel oil from the sunken ship started leaking and started to be an environmental hazard. The rate of leakage was estimated at around 10 tons per day and this oil was being washed to the shores of Mumbai.

This disaster in the Arabian sea is reminiscent of a similar disaster in August 2010, when the two Panamanian carriers MSC Chitra and MC Khalijia collided near the Mumbai port. This collision was due to a communication failure between the two ships who were operating at different frequencies. This collision resulted in a spillage of oil as well as a lot of pesticides (which was the cargo).

These disasters are also reminiscent of the Exxon Valdez oil spill when an oil tanker bound for California struck Prince William Sound's Bligh Reef and spilled 300,000 barrels of crude oil.

The effects of these disasters included:

1. It caused the death of fishes as the oil entered their gills and hampered in their respiration. The pesticides and oil when ingested could make the fish inedible.
2. Fishing in the affected areas was banned and this affected the livelihood of fishermen.
3. The operations in the port were hampered and this caused a significant economic loss.

The location of such accidents and the timing are important. The nuclear reactors in Mumbai use sea water for their cooling operations. The oil spill in the sea near the reactors implies that these reactors have to take additional measures to purify the sea water before using it as this may affect their research activities.

Also, the ecology of the mangroves along the coastal regions is affected. If the spill occurs during the season of flowering of mangroves the seeds will be coated with a layer of oil. This, in turn, will hamper their germination and the ecosystem would be affected.

The case studies done so far give an overview of how anthropogenic activities affect the environment. In most cases, it is the indiscriminate use of toxic chemicals (DDT, endosulfan). In other cases, an anthropogenic activity at one level can result in adverse impact on another part of the ecosystem (diclofenac) and in other cases it could be a desire to grow rapidly at the expense of the environment (Coca-Cola in Kerala). The case studies discussed serve to increase awareness about different issues which can affect the environment.

Ethics

Many companies are reputed for their highly ethical behaviour and there are several others who are notorious because of their unethical practices. Companies like Enron, for instance, have not been successful because of their unethical practices. These usually arise because there is high pressure on the top brass or management to meet unrealistic demands. Unethical practices may be beneficial in the short term but will hurt the company in the long run.

The practising engineer or professional comes across many situations in his career where there is a conflict of interest. The right decision is not clear. In such situations there are several possible alternatives to a situation. The decisions taken are determined by several factors including cultural background and the solutions are not unique. In such situations which the professional faces often in his career, there are several options or lines of actions he can follow. He has to choose the "right one". The objective here is to make an informed decision. What may be right or ethical to one may be unethical to another. While there is no clear-cut methodology as to what is ethical or unethical it is useful to:

1. Identify the source of the conflict of interest.
2. List what has a significant bearing on the decision.
3. Identify people who would be adversely or favourably affected by the decision.
4. Put yourself in the shoes of the affected persons and ask yourself what would their expectations be.
5. List different alternative plans of actions and their effects.
6. Decide on a final course of action.

To expose the student to such situations different case studies are used and discussed in the class. The case study is given a priori to the class and various options that can be possible, choices are listed. The students are then asked to come individually and present how they would respond to a situation. A fairly good distribution of choices made by the students usually emerges in such a situation. We now present the following case studies. Several of them are taken from Shallcross (2003)*.

Case Study 7

The conflict in a mentor. Gurumurthy (Guru) is a senior colleague in an organization, Overful. He is a mentor for many of the younger colleagues in his organization. Chellakumar (Chela) is one such sincere, hard working young

---

* Reprinted with permission from Dr. David Shallcross, Associate Professor and Head, Department of Chemical and Biomolecular Engineering, Melbourne School of Engineering, University of Melbourne, Victoria, Australia.

person in the organization and has been around for five years working with it. He has always looked up to Guru for good advice on the professional front. However, in Overful there is not much scope for career growth as all the positions above Chela are occupied. He does not see any changes in this in the near future. However hard he works, he is not likely to see a significant rise in his position in the company nor in his salary.

Chela has found another organization "Underfull" which is looking for people with the kind of experience he has. The pay package in this new organization is also highly attractive. However, his application has to be accompanied by three letters of reference and he approaches Guru for one of these. Guru is a natural choice since he knows the work of Chela very well.

Imagine that you are Guru. What would be your course of action?

1. You would give a clean reference letter to Chela although this would mean your company losing a sincere and hard working person who has contributed to the growth of the company in the past.
2. You would refuse to give a letter of recommendation although this would keep him in the company but you would be acting as a bad mentor. Possibly, Chela would have a bad relationship with you and may not be motivated after this to do any work.
3. You should write a negative recommendation letter and not inform Chela about it, this way Chela will not leave but continue in your organization and he would not know why he was not selected for the new position.
4. You will talk to Chela about not leaving Overful and advise him to continue working there.
5. You will talk to higher management in Overful and convince them to give an attractive offer to Chela to dissuade him from having thoughts of leaving the company.
6. You will talk to Chela about leaving Overfull.

Case Study 8

Ramesh, a young engineer, has just been moved to a plant recently acquired by his company. He is one of the four plant superintendents. Ramesh is told by his group's vice-president to improve the plant's compliance with government regulations and to reduce the plant's high accident rate. Unfortunately, Karthik, the plant manager and Ramesh's immediate superior does not want to change any of the plant's operations in his last years before retirement. Of particular concern to Ramesh are the plant's five chlorobenzene reactors which, for safety reasons must be operated at temperatures below 160°C. Ramesh believes that a culture of unsafe practice has grown in the company with operators not following written procedures but instead neglecting safety procedures with the connivance of unit and shift supervisors. Ramesh's proposal to spend money on improved control equipment is rejected by Karthik and Shankar, the group's business managers. Later on one of the reactors shows a "runaway" resulting in an unplanned discharge to the environment. In response to the incident Ramesh is directed to fire or demote the operator on duty at the time. Ramesh disagrees with the action as it is unfair and does not address the real cause of the problem. These circumstances cause Ramesh to consider the course of action he should follow. What should he do?

Case Study 9

Mary is the chief process engineer for the ABC chemical company. She works at the company's plant in a small town. Indeed, the company is one of the few remaining major employers in the area and the town's economic future depends heavily on the health of the plant. Recently, Mary has been given the new responsibility for environmental compliance at the local plant level. Mary takes her new responsibilities seriously and spends times attending seminars and workshops on environmental compliance. At one session dealing with the clean-up of contaminated groundwater at industrial facilities, she learns of a number of actual case studies. She is struck by the similarities between some of the case studies and her own plant and begins to suspect that there may be a major problem at her own site. She believes that it may be possible that contamination from the plant's facilities over the years may have escaped into the town's water supplies. She fears that the drinking water might be contaminated. She has no proof that there is contamination, yet the very act of investigating the situation may lead to the closure of the plant with a loss of many jobs. Conversely, if she does nothing and a subsequent investigation will reveal that the drinking water is contaminated and that she knew but did not take action, she would be in a very difficult situation. What should she do?

Case Study 10

For nine years, Binh worked for Canberra Petrochemical Company, CPC, before joining MUCESS chemicals. While talking to Jim, an old friend from CPC he learns that one of his former colleagues from CPC, Sharon, an engineer in the maintenance department, had left her job. Binh had dated Sharon several times five years earlier and was aware that she was not well liked by some of her colleagues. Jim mentions to Binh that just before she left there had been an incident during a routine maintenance job in a reactor which resulted in considerable damage but no injuries. Jim passes on to Binh the rumour that it was Sharon's fault and that she was about to be fired before she resigned. Knowing Sharon as he does, Binh could well believe that the incident could have been her fault. Now Binh learns that Sharon has applied to work at MUCESS chemicals in the maintenance group. While Binh does not work directly in that group, he is approached by Fred, maintenance head, and asked for his opinion of Sharon. In the course of the conversation he learns from Fred that Sharon has been given good reference from her former supervisors at CPC and that the reactor maintenance incident has not been mentioned. What should Binh do? Should he alert Fred to what he has heard about Sharon's role in the accident? Should he arrange to meet Sharon and ask her directly about the accident? Should he decline to comment on the grounds that he once dated Sharon? Should he just do nothing at all?

Case Study 11

Joan had worked as a production engineer at Borkon Corporation's polymer plant for six years. Her responsibilities related to the production of a high value, speciality polymer, PMRT. By changing the type of impeller used in the reactor and by adjusting how the raw materials are fed into the reactor she was able to improve the process by about 15%. Then, after a six-month development program she was able to achieve an additional 5% improvement by changing the location and shapes of the baffles. Much of the work associated with the baffle project was done in her own time at home on her personal computer. Joan's supervisor is Ted, a veteran of the plant who has a habit of taking credit for other people's contributions. In a difficult market Borkon decides to shed staff and Joan is told that she is to leave the company. She knows that Ted has not been happy with her attempts to get recognition for her work which Ted has denied her. After 4 months of unemployment Joan lands a job with Polymasters, a rival to Borkon. Her new responsibilities include the production of PMRT. She quickly realizes that Polymasters still uses the same sort of reactor configuration that Borkon did before the extensive modifications were made. Which, if any, of the following should Joan consider to be confidential information? The use of the better, commercially available impellers, the computer program she developed at home and/or the revised location and shape of the baffles. How should Joan go about improving the yields of the PMRT reactors? Should she immediately switch the impellers and change the way in which the raw materials are fed into the reactor? Should she immediately change the baffle locations? Should she discuss her situation with her new company's legal staff? Should she not make any changes at all?

Case Study 12

Barry has been working as a Process Engineer at an oil refinery in the western suburbs ever since graduating from University four years ago. One Saturday morning at the Victoria Market Barry bumps into Finn, a friend from his university days whom he has not seen for three-and-a-half years. After chatting briefly they arrange to meet for lunch on one of Barry's days off a week later in the city near Finn's office.

Over lunch Barry and Finn talk about the old days at university and about their respective jobs. Barry learns that Finn has been promoted to Inspector with the Environmental Safeguard Agency, a government agency. Finn's job is to investigate allegations of water and air pollution from industrial sites in the south-eastern suburbs of Melbourne. Finn describes his powers under the law to enter sites looking for violations of site discharge licences. Barry then describes his duties as a Process Engineer within his company. Barry is aware that his company doesn't always obey the letter of the law with regard to discharges and so is careful about what he says to his friend knowing that a careless comment could be damaging to his employer. During the conversation Barry admits his ignorance of the details of the law so Finn invites Barry up to his office to give him some printed booklets prepared by the ESA.

As Finn searches through his files for the booklets in his office Barry glances around the cluttered room and notices a file labeled 'CITOX Pvt Ltd' on the top of the desk. As he hands Barry a copy of the booklet, Finn notices his friends gaze and casually mentions that in two days time he will be leading a raid on Citox in Oakleigh. Finn tells Barry that the ESA have received a tip-off that Citox have been exceeding their discharge licences and that on his first raid as an inspector he hopes to get evidence for a prosecution in the courts.

As Finn tells him this, Barry listens quietly, his mind racing. Barry has not told Finn that his girlfriend, Eliza is a chemical engineer working at Citox's Oakleigh site. Only the other night Eliza had revealed that she had had to authorize the shift supervisor to partially vent the contents of a reactor vessel to the atmosphere after the reactor had been incorrectly filled.

Finn has given Barry information about a secret raid unaware that Barry's girlfriend works at the site. What should Barry do with this information?

- A. Do nothing. Barry was given information that he should not have been. He did not ask to be told of the impending raid. Barry can pretend that he was never told and not tell Eliza. But if Eliza were to ever to find out that Barry could have warned her about the raid the knowledge could severely strain their relationship. Also if Finn were ever to learn that Barry's girlfriend worked for Citox and Barry had not told him then Finn could feel, however unjustified, that he had been betrayed by Barry.

- B. Tell Finn about Eliza and her association with Citox. This action could force Finn to cancel the raid, and could place Finn in an embarrassing and difficult position within the ESA. Finn could be reprimanded for revealing confidential information and the incident could effect his career. Alternatively, Finn might stand firm and continue with plans for the raid. He might caution that any attempt to warn Eliza or any other Citox employees of the impending raid could have serious repercussions for Barry.

C. Warn Eliza without revealing his source. Barry could mention to Eliza that he had heard a rumour that Citox were about to be inspected by the ESA. The danger is that Eliza might want to know the source of the information and might not act on it if Barry does not reveal his contact. Also, Finn could realize that Citox had been tipped off by someone and may guess that it was Barry.

What would you do?

Exercises

1. Material Safety Data Sheet (MSDS) for different chemicals contain information on toxicity. Choose a chemical and find its MSDS. This will increase your awareness on how to determine if the chemicals you would handle later on in experimental laboratories are safe or unsafe. This will help you understand the precautions to be taken while carrying out experiments and how to handle spills in laboratories, etc.

2. The use of mosquito repellants is very common in households. Choose any one brand and find out how it works? What chemicals do they use? Are these safe for human beings? Could it be that some of these can have an adverse effect on indoor air quality and human health?

3. You are a final-year student, (a senior) who is about to graduate. There is a recession in the economy and the companies are not recruiting people. You are desperately seeking a job and you approach one of the professors to help you with his industrial contacts. The professor helps you get a good job in company A. You are supposed to start working in August. You graduate from the college in June and in July you get a better offer with a higher salary from another company B. What would you do?

    - You will say no to company B on the grounds that your professor helped you get a job and you may end up spoiling the relationship of your Professor with company A? Accepting company B's offer will also mean that your Profesor may not help students in a similar situation in the future.
    - You will accept company B's offer and send a regret letter to company A and inform your professor.
    - You will accept company B's offer and send a regret letter to company A and not inform your professor.
    - You will discuss this with your professor and abide by his suggestion? Will you join company A and work for two years or so and then shift possibly to company B?
    What would your response be if you have already joined company A and then the offer from company B comes?

4. You are a final-year student, who is about to graduate. You are interested in going to a top school in the United Statees for higher studies. All your friends have offers and they are all going and you are left behind. You are desperately seeking an admission with financial support and you approach one of the professors to help you with his university contacts. The professor helps you get an offer from university A. You are supposed to join the graduate school in August. After the professor makes this arrangement you get an offer from university B which is ranked higher with financial support. What would you do?

    - You will say no to university B on the grounds that your professor helped you get the admission to university A and you may end up spoiling the relationship of your professor with university A. Accepting university B's offer will also mean that your professor may not help students in a similar situation in the future.
    - You will accept university B's offer and send a regret letter to university A and inform your professor of your decision.
    - You will accept university B's offer and send a regret letter to university A and not inform your professor of your decision.
    - You will discuss this with your professor and abide by his suggestion.

5. You are an administrative assistant in the human resources department. Your good friend Sheela is applying for a job with the company and you have agreed to serve as a reference. Sheela approaches you for advice on preparing for the interview. Being in the HR section you have the list of actual interview questions asked of applicants. What will you do?

6. You work in quality control department of your organization. Once a year, your company gives away the old computers to the local orphanage. No specific records are kept of this type of transaction. You really need a computer for your son who is in college. Your supervisor asks you to deliver computer systems to the orphanage. What would you do with the computers?

7. Shirley was recently hired to work as a receptionist for the front lobby. As a receptionist, she is responsible for taking printout for the associates. Her son, Jason, comes in and needs a printout for a school project. He brought his own paper and needs 200 pages to be printed. If he doesn't bring the printout with him, he will fail the project. There is no security key for the printer and the company does not keep track of prints made by departments. Would you take the printout?

References

1. *EFSA Journal*, Statement on toxicity of endosulfan on fish.
2. International Program on Chemical Safety (IPCS), *Health and Safety Guide*, No. 17.

3. EXTOXNET (Extension Toxicology Network), Pesticide Information Profiles.
4. State of Endosulfan, CSE WEBNET.
5. www.ban endosulfan_join the campaign.htm.
6. Endosulfan victims, a division of Kasaragod Vartha.
7. NCBI, Endosulfan: a clinical profile.
8. Health Canada, www.hc-sc.gc.ca.
9. Roger E. Meiners and Andrew P. Morriss, Pesticides and Property Rights.
10. Shallcross, D., "Experiences Teaching Professional Ethics to Chemical Engineering Students", *International Conference on Engineering Education*, 2003, Valencia, Spain.
11. Shallcross D.C. and Parkinson M.J., Teaching Ethics to Chemical Engineers, Some Classroom Scenarios, Transactions IChemE, PartD, 2006, Education for Chemical Engineers, v1, 49–54 (www.icheme.org/ece, doi:10.1205/ece/05011)

http://panchabuta.com/2011/04/11/andhra-lake-wind-power-project-of-enercon-india-under-scrutiny-of-ecology-panel/

http://www.thehindu.com/news/states/other-states/article1682920.ece

http://www.hindu.com/2011/04/10/stories/2011041052212200.htm

http://www.thehindu.com/todays-paper/tp-national/article1698234.ece

APPENDIX

MATLAB Programs

The application of the fundamental laws of conservation usually leads to equations of different levels of complexity. These could be algebraic in nature or they could be differential equations. Engineers must, hence, have a sound knowledge of mathematics to analyse these equations. The properties of different equations and their methods of solution are discussed in basic mathematical courses. These courses equip the engineer with the details required for a complete mathematical problem formulation.

The engineer must know how to solve these equations. On most occasions the solution can be found only after a computer program is required. The chemical engineer uses Fortran as the basic scientific language. Several software packages such as Aspen Plus are written in this package.

MATLAB is powerful software which enables computations to be performed accurately and elegantly. This is a commercial package and an analogous open source package (hence completely free) Scilab with same features as MATLAB is available. Moreover, in these packages results can be depicted graphically with ease.

These packages have tools containing algorithms written by professionals trained in the art of writing programs to solve different kinds of problems which arise in engineering. These are general-purpose tools and can be used to solve a general system of equations. The engineer should know how to use these tools effectively. He should not spend his time writing a "better code" but in using the tools to get results for applications.

A good engineer must know the basics of numerical algorithms to use these packages effectively. This will enable him to determine if the results generated by the program or software package are accurate. It will also help him detect errors and make corrections in the code when the results are found

168 Appendix—MATLAB Programs

to be inaccurate. The engineer must know how to validate the results and not blindly accept the results coming from packages.

In addition to the above packages several commercial software packages exist which facilitate computations. Students must desist from using these as black boxes. These packages are elegant tools but the effectiveness of these tools is determined by the knowledge of the user.

To help the student get started with programming in MATLAB some simple programs are given below. These programs help solve the equations discussed in the text and generate the results presented there.

MATPROG 1

This contains a program in MATLAB which solves for the best fit R using equation 4.9.

A program in Matlab which finds R_{mean} and $R_{leastsq}$ based on equation 4.9

```
%A program in Matlab which finds Rmean and Rleastsq based on
equation 4.9

i=1:1:10;

v=[15,23,36,39,54.5,58,73.9,83,87,105]; %defining voltage values

R_avg=mean(v./i); % calculating mean of R

v1=R_avg*i; %voltage predictions using R_avg

R_least=v/i; % R calculated using least squares method

v2=R_least*i; %voltage predicted using least squares method

plot(i,v,'o',i,v1,i,v2,'-'); %Plotting the velocity and current
using the two methods

xlabel('Current, i(amp)','fontweight','b')

ylabel('Voltage, V(volt)','fontweight','b')

legend('Expt data','Rmean','Rleast square','location','NW')

%------------------
% R_avg  =10.9915
% R_least =10.2852
```

MATPROG 2

A program in MATLAB which solves equation the ordinary differential equation 4.17.

```
function tank_height ()
global qin alpha A
% Flowrate of water into the tank (m3/min)
qin=10;
% Cross-sectional area of the tank (m2)
A=0.785;
%The outlet flow parameter alpha (m^2.5 min^-1)(Eq. 4.17)
alpha=8;
% The steady state height when the rate of change of height is zero
hss=(qin/alpha)^2;
% The deviation (dev) of the initial height from the ultimate steady state
dev=.6;
% the time span for which the integration is performed (10 min)
tspan=[0,2];
% Initial condition of height. Add or subtract dev from hss depending on
% whether the tank height is initially above or below the steady state value
ho=hss+dev;
% solving the Diff eq. using the 4th order Runge Kutta based solver in Matlab
[T,H]=ode45(@fun,tspan,ho);
% plotting the results
plot(T,H,'linewidth',2);
axis([0 1.5 0.8 2.5])
text(0,hss,'———————————————')
text(.05,hss-.1,'Steady State','fontsize',10);
xlabel('time (min)','fontsize',12,'FontWeight','bold')
ylabel('height(m)','fontsize',12,'FontWeight','bold')
end

%The function which returns the RHS of the differential equation
function dh= fun(t,h)
global qin alpha A
dh=1/A*(qin-alpha*(h)^.5);
end
```

MATPROG 3

Matlab program to plot dependency of conversion on Damkohler number.

```
% creating an array of damkohler nos. (d) from 0 to 100 with
an interval of 0.001
>> d=0:0.001:100;

>> plot(d,d./(1+d))
```

MATPROG 4

Matlab program for finding dependency of concentration on flow rate

```
v=[1 10 50 100]; %Defining various reactor volumes
q=0:1:100; % Defining a array of 100 flow rates
c=zeros(1,100); %Allocating a array of 100 elements to store
computed conc
cf=10; % Feed Concnetration
k=2; % reaction rate constant
% computing the concentration vs flowrate profiles for each
reactor volume
for i=1:4
  c=cf.*q./(q+v(i)*k);
  plot (q,c)
  hold on % To plot all profiles are plotted on the same graph
end
xlabel('q','fontweight','b')
ylabel('C_A','fontweight','b')
hold off
% indicates that all profiles are plotted. (the next plot
command
%will plot a new graph)
```

Index

Air liquefaction unit, 45
Alkali Act of 1863, 55
Antoine equation, 97
API gravity, 12
Atmospheric pollution, 15
Autism, 155

Batch processing, 27
Batch reactor, 70
Best possible fit, 80
Bharat stage norms, 15
Bio-accumulate, 147
Boiling points, 56
Boundary condition, 103
Breast cancer, 155
Brownian motion, 129

Calculus of variations, 39
Carcinogenic, 146
Catalytic
   cracking, 9
   reactions, 67
   reforming unit, 11
Centralized treatment plants, 5
Centrifugal pump, 60
Chemical mass balance, 78
Chemical reaction engineering, 69
Chemical vapour deposition (CVD), 19
Clausius–Clapeyron equation, 97
Closed systems, 74, 82, 83
Closure problem, 133
Coal gasification, 13
Coca-Cola, 156
Coke deposition, 10

Computation techniques, 97
Condensers, 64
Contact process, 39
Continuous processing, 27
Continuous stirred tank reactor, 70
Control
   mass, 84
   surface, 84
   volume, 84
Converging pipe, 86
Conversion, 123
Coupled reactor-separator, 46
Cracking, 9
Crude oil, 7
Cryptorchidism, 154

Damkohler number, 123
Darcy friction factor, 121
DCDA process, 41
DDT, 145
Deforestation, 150
Degrees of freedom, 96
Delayed coker unit, 11
Desalination, 48
Diclofenac, 143
Diesel, 8
Dimensionless analysis, 111, 112
Dimensionless numbers, 111
Disinfection, 49
Dispersion, 129, 133
   of pollutants, 127
Distillation, 7, 56
Distributed systems, 100
Downstream operations, 6
Drag force, 111
Dry etching, 2

Economics of scale, 109
Economy of scale, 4
Eddy viscosity, 135
Endocrine, 155
  disruption, 146
  disruptor, 154
Endosulfan, 151
Equation of continuity, 93
Equilibrium compositions, 59
Ethics, 159
Euler's acceleration formula, 86
Eulerian derivative, 86
Euro norms, 15
Evaporators, 64
Extract, 59
Extraction, 66

Fabs, 21
Feral dogs, 143
Fick's law, 68
Filtration, 48
First law of thermodynamics, 83
Fischer–Tropsch, 13
Flowsheet, 10
Fluid catalytic cracking unit, 11
Fluidized bed reactor, 10
Fluidized catalytic cracker, 10
Fourier's law, 68
Friction factor, 117

Gas turbine, 14
Gay–Lussac, 35
Genotoxicity, 146
Germanium, 19
Glover tower, 37
Greywater, 6

Heat exchangers, 4
Heat integration, 4
Heat transfer, 7, 61
  coefficient, 63, 102
Heavy distillates, 9
Heterogeneous reaction, 67
Hydro cracking, 9
Hydrocephalus, 152
Hydro-desulphurization, 6, 11

Indoor air quality, 2
Initial condition, 103
Integrated circuit (IC), 18
Integrated gasification combined cycle
  (IGCC), 3, 14
Ion implantation, 20
Isolated systems, 83

Kerosene, 8
Kerosene-based stoves, 2
Kirchhoff's first law, 92

Lagrangian derivative, 86
Laminar, 117, 119
Laws of
  conservation of energy, 74
  conservation of mass, 74, 79
  conservation of momentum, 74
LeChatelier's principle, 39
Lead chamber process, 34
Least squares error, 77
Leblanc process, 43
Linear algebraic equations, 76
Liquefied petroleum gas (LPG), 2
Liquid nitrogen plant, 52
Local derivative, 86
Lumped systems, 100

Malaria, 146
Maleic anhydride, 71
Material derivative, 86
Method of least squares, 79
Microchip, 18
Micro-gravity conditions, 3
Minimum error, 76
Mixing, 3
Modeling, 99
Momentum transfer, 60

Natural convection, 129
Newton's second law, 94
  of motion, 83
Newtonian fluid, 68
Nikuradse equation, 118
Nobel prize, 145

Nuclear reactions, 142
Numerical methods, 97
Nusselt number, 125

Ohms' law, 68, 81
Open systems, 74, 82, 83, 84
Optimal temperature profile, 39
Osmosis, 23
Osmotic pressure, 49
Over determined system, 96

Packed bed reactors, 9
Pain killer, 143
Paint manufacture, 28, 31
Particulate matter, 13, 78
Persistent organic pollutant, 147
Petroleum
    coke, 11
    refining, 7
Phytoremediation, 6
Pilot plant, 3, 109
Plug flow reactor, 70
Plumes, 129
PM_{10}, 78
Pressure drop, 112
Pressure swing adsorption, 52
Principle of heterogeneity, 114
Process engineer, 3
Product engineer, 3

Rabies, 143
Raffinate, 59
Receptor modeling, 78
Reciprocating pump, 60
Recycle streams, 45
Refinery, 7
Reliance refinery, 33
Renewable energy, 149
Residence time, 69
Reverse osmosis, 46
Reynolds number, 117
Reynolds transport theorem, 87, 95
Runaway, 108
    condition, 69

Scaling down, 4
Sedimentation, 48

Semiconductor processing, 18
Semi-empirical, 128, 129
Series parallel network of reactions, 71
Settling velocity, 112
SHE (Safety, Health and Environment), 137
Shutdown, 41
Silicon, 19
Similarity of triangles, 112
Simulation, 99
Soda ash, 43
Sodium carbonate, 43
Solvay process, 43
Sour crude, 11
Source apportionment problem, 77
Spin-on coating, 19
Stability, 91
Stable-steady states, 91
Start-up, 41
Steam turbine, 14
Stirrers, 60
Sudden contraction, 92
Sulphuric acid manufacture, 34
Sustainable, 150
Synthesis gas, 13

Thermal cracking, 9
Thermodynamics, 59
Three-way catalytic convertor (TWC), 18
Transport phenomena, 57, 68
Turbulent, 117
    flows, 127
    viscosity, 135

Under determined system, 96
    operations, 56, 57
    processes, 57
Upstream operations, 6

Viscosity acts, 7

Water gas shift reaction, 14
Wax crystallization, 7

Xenoestrogen, 154